Do It Yourself

This book is dedicated to people everywhere who are changing their worlds, and creating everyday revolutions by doing it themselves – loving, laughing, living, through solidarity, resistance and defiance, for a more just and ecologically sound world.

Do It Yourself

A Handbook for Changing our World

Edited by
The Trapese Collective

www.handbookforchange.org

Pluto Press
LONDON • ANN ARBOR, MI

First published 2007 by Pluto Press
345 Archway Road, London N6 5AA
and 839 Greene Street, Ann Arbor, MI 48106

www.plutobooks.com

British Library Cataloguing in Publication Data
A catalogue record for this book is available from the British Library

Hardback
ISBN-13   978 0 7453 2638 2
ISBN-10   0 7453 2638 2

Paperback
ISBN-13   978 0 7453 2637 5
ISBN-10   0 7453 2637 4

Library of Congress Cataloging in Publication Data applied for

10   9   8   7   6   5   4   3   2

This book is printed on paper accredited by the Forest Stewardship Council and is
suitable for recycling and made from fully managed and sustained forest sources.
Logging, pulping and manufacturing processes are expected to conform to the
environmental regulations of the country of origin.

Designed and produced for Pluto Press by
Chase Publishing Services Ltd, Fortescue, Sidmouth, EX10 9QG, England
Typeset from disk by Stanford DTP Services, Northampton
Printed and bound in the European Union by
CPI Antony Rowe, Chippenham and Eastbourne

contents

illustrations

The Trapese Collective is Alice Cutler, Kim Bryan and Paul Chatterton.

After studying Social Anthropology, living in the Can Masdeu community in Barcelona, and campaigning with London Rising Tide, Alice has recently settled back in Brighton where she is putting her ideas into practice – running workshops, teaching English to asylum seekers, community gardening, writing and cooking at the Cowley Club social centre.

Kim travels between Escanda (Spain), Dublin and Trapese (UK) trying to avoid getting tied up, but simultaneously wanting to settle down – juggling love, life, jobs, politics, worrying about ensuing climate chaos and growing vegetables.

Paul can normally be found teaching and researching on autonomy and international politics at Leeds University. His other lives include helping to set up a social centre, the Common Place, and a housing co-operative in Leeds, volunteering with Kiptik, a Zapatista solidarity group and spending time on his allotment. Previously, he has written *Urban Nightscapes: Pleasure Spaces and Corporate Power* (2003) and *Taking Back Control: A journey through Argentina's popular uprising* (2004).

acknowledgements

In this book, we have tried to gather into one place many of the inspiring examples and ideas that we have come across in an attempt to make them more accessible and possible to realise. We see this as one part of our ongoing work to communicate these analyses and initiatives with as many people as we can. This book would not, of course, exist without the thousands of people who work away on these projects, campaigns and networks. It would not exist either without all the people who have helped us organise and who have come to our workshops over the years, asked questions, engaged with these debates and given us the motivation to write it.

Our inspirations come from the many movements and people within them who are acting to make the world a more sustainable, fairer place. We would particularly like to thank everyone involved with: the Common Place, the Cowley Club, the Sumac Centre, Escanda, Can Masdeu, Seoma Sprai, and all the other radical social centres, Rising Tide, Schnews, Indymedia, Kiptik and the Zapatistas, No Borders Network, Clearer Channel, the Wombles, Seeds for Change, the Dissent! Network, Shell to Sea and Rossport Solidarity Camp, the Piqueteros, the Camp for Climate Action, CIRCA, Smash EDO, Corporate Watch, Café Rebelde, the Permaculture Association, PGA, (People's Global Action), Activist Trauma Network, Via Campesina, Scottish Education for Action and Development, Cre8 Summit, Bristle, the Aubonne Bridge Campaign, Carbon Trade Watch, International Solidarity Movement, Anarchists Against the Wall, Earth First!, Food Not Bombs, Moulsecoomb Forest Garden, Xanadu and Cornerstone housing co-operatives in Leeds, EYFA, Mama Cash, Carbusters, Radical Routes, the Advisory Service for Squatters, Bicycology, and the many struggles against privatisation in the UK and beyond.

There are also many specific people whose contributions and advice have made it possible to produce this book, and thanks to you all. Especially, we are very grateful to the chapter authors who had the time, patience and confidence in us to contribute their work to this publication; Andy Goldring, Bryce Gilroy-Scott, Seeds for Change, Tash Gordon and Becs Griffiths, Jennifer Verson, the Vacuum Cleaner, Stuart Hodkinson, Chekov Feeney, and Mick Fuzz. Also to the many people and groups who have added their experiences and stories that make this book what it is, including; Starhawk, George Marshall, Earth Haven, Malamo Korbetis, Jo and Tony

Bryan, Warren Carter, Donna Armstrong, Ruth O'Brien, Graham Burnett, Insight Video, the Porkbolter, Mark B, Ziggy, Dave Morris, Fuzz, Freya, Anarchist 606, Isy, Claire Fauset, John Jordan, the Alberta Council for Global Co-operation and Nicola Montanga.

A very big thanks to everyone that has helped make the book look as good as it does, namely; UHC Design Collective for the cover of the paperback edition, Andrew X for suggestions for the page design and all his other help, Ade Lovejoy and Alex Mac for the chapter icons, Simon Liquoricefish, Guy Pickford and Alison at Leeds University for the illustrations, Guy Smallman and Active Stills for photographs. Also thanks to Seth Wells, (Vector Pixel) for the website and Brian for the leaflet.

Thanks also to Melanie Patrick and Robert Webb of Pluto Press for bringing it all together and David Castle for his supportive role as commissioning editor at Pluto Press, especially for his backing of Creative Commons publishing. Also thanks to the School of Geography at Leeds University for their support.

Last but certainly not least, a huge thanks to all our families, partners, compañeros and friends who have been such a great help and support to us. To anyone we have forgotten to name, thanks.

Putting this book together has been a very educational and inspirational experience for us. We worked collectively throughout, meaning that at times things took a lot longer than if someone had just taken charge, but we were anxious that the book was a reflection of our politics and desires for change. We made some mistakes, fairly inevitably, mainly because we had not done anything quite like this book before, but we hope that the end result conveys our excitement about the ideas and solutions that are presented. In a book like this there are sure to be omissions. Please feel free to add comments, updates, corrections, and extra resources to the book website by emailing us on address below.

Alice, Kim and Paul
trapese@riseup.net
www.handbookforchange.org

glossary

Affinity group   An affinity group is a small group of activists (3–20) who work together on direct action. Affinity groups organise using non-hierarchy and consensus. They are often made up of friends or like-minded people and provide a method of organisation that is responsive, flexible and decentralised.

Anarchism   Anarchism is a political philosophy and a way of organising society; it is derived from the Greek – without rulers. It is a belief that people can manage their own lives, and so rulers are undesirable and should be abolished. For many anarchists, this also includes institutions of authority, such as the state and capitalism.

Appropriate technology   Appropriate technology questions excessive technology and the problems of industrialism and has mainly been used in developing nations or underdeveloped rural areas. It uses the simplest and most benign technologies and responds to community need rather than those of the state or private interests.

Autonomy   Stemming from the Greek, meaning 'self-legislation', autonomy is a belief system that values freedom from external authority. This can occur at the individual and collective level. Autonomy has widespread use for many contemporary social movements trying to manage their own lives and communities.

Commons   Traditional use of commons referred to traditional rights such as animal grazing. More recently, commons refers to common rights in a community for other resources and public goods, such as water, oil, medicinal plants and intellectual knowledge. Many social movements are struggling for these resources to remain in common ownership.

Composting   Compost is recycled organic matter, especially kitchen and garden waste. Compost bins allow this organic matter to quickly decompose. Composting is a popular way to reuse household waste in order to produce usable compost for the garden and for growing food. Composting also reduces landfill (a major source of greenhouse gases).

Consensus   Consensus is a way of making decisions that aims to include everyone in the decision making process and resolve any objections. It is a form of grassroots or direct democracy and rejects representative forms of democracy associated with voting and hierarchy which can ignore the views of minorities.

Direct action   Direct action is a form of political activism which rejects reformist politics such as electing representatives as ineffective in bringing about change. It involves us taking responsibility for solving problems and achieving demands using strikes, occupations, blockades and other forms of public protest.

Do it yourself culture   A broad term referring to a range of grassroots political activism with a commitment to an economy of mutual aid, co-operation, non-commodification of art, appropriation of digital and communication technologies, and alternative technologies such as biodiesel. DIY culture became a recognised movement in the 1990s in the UK, made famous by direct action and free party culture.

Enclosure   Enclosure is the process of subdividing communally held land for individual ownership, mainly associated with twelfth- to nineteenth-century England. Contemporary movements against the privatisation of land and the sale of public goods are regarded as struggles against the new enclosures.

Hacklab   A hacklab, or media hacklab, is an autonomous technology zone used for the promotion, use and development of emancipatory technologies such as free software and alternative media. Hacklabs promote active participation and creative use of technology.

Hierarchy   A hierarchy is a system of ranking and organising things or people, where each element of the system is subordinate. Many inequalities in our society stem from the fact that most social organisations, such as businesses, churches, workplaces, armies and political movements, are hierarchical.

Indymedia   Indymedia means Independent Media. It also refers to a global network of independent journalists and alternative media which use an open source publishing model to empower people to bypass corporate media. Started in 1999, there are now over 190 Indymedia outlets around the world.

Movements of movements This term is used to describe loose groupings often called the global justice or the alter/anti-globalisation movement. It refers to large gatherings, like the World Social Forums, and large convergences at global summits since protests in Seattle, 1999. It emphasises the diversity of aims and tactics of participants and challenges Marxist-Leninist organisations who call for a unified programme for social change.

Mutual aid Mutual aid describes a principle central to libertarian socialism or anarchism, and signifies the economic concept of voluntary reciprocal exchange of resources and services for mutual benefit. It is a key idea within the anarcho-communist, co-operative and trade union movements.

Peak oil Also known as Hubbert's peak, peak oil refers to the peak of the entire planet's oil production. After peak oil, where predictions range from the present to 2025, the rate of available oil on Earth will enter a terminal decline every year. It signifies massive changes for our oil dependent societies.

Permaculture Coined by Australians Bill Mollison and David Holmgren during the 1970s, permaculture is a contraction of permanent agriculture. It is an ethical design system intended to be sustainable and applicable to food production, land use and community building.

Popular education Popular education is an educational process designed to raise the consciousness of its participants and allow them to become more aware of how an individual's personal experiences are connected to larger societal problems. Participants are empowered to act to effect change on the problems that affect them.

Revolution A revolution is usually a drastic, rapid change in social or political systems, or a major change in a society's culture or economy. It is often associated with violence and seizing state power, but it can also refer to incremental radical change at the everyday level.

Squatting Squatting is the act of reclaiming and occupying abandoned or unoccupied spaces that the squatters do not own, rent or otherwise have permission to use. It includes land squats, shanty towns, homes, social centres, gardens or protest sites.

Social centres   Social centres are community spaces used for a range of not-for-profit activities, such as support networks for prisoners or refugees, cafes, free shops, public computer labs, meetings and gatherings, art projects and benefits advice. They are managed collectively by participants and can be rented, bought or squatted.

Solidarity   Solidarity refers to the feeling of unity based on common goals, interests and sympathies. It is a term which is promoted by many social movements to help create social relationships based on justice and equality.

Spokescouncil   A spokescouncil is a collection of affinity groups who meet together for a common purpose. It is named after the 'spokes' of a bicycle wheel. Each spokesperson represents the affinity group in the spokescouncil which makes decisions via consensus.

Subvertising   Subvertising makes spoofs or parodies of corporate and political advertisements. This can take the form of a new image or an alteration to an existing one. A subvertisement can also be referred to as meme hack, social hacking or culture jamming.

Sustainable development   Sustainable development aims to meet the needs of the present without compromising the ability of future generations to meet their own needs. It aims to overcome environmental degradation without forgoing the needs of economic development, social equality and justice.

Glossary extracted and adapted from www.wikipedia.org, an open-source free encyclopedia.

introduction   do it yourself

The Trapese Collective

Everyday, everywhere, through spontaneous and planned actions, people are changing the world, together. These everyday actions come from the growing desire to do it ourselves – plant vegetables, organise a community day to get people involved in improving where we live, expose exploitative firms, take responsibility for our health, make cups of tea in a social centre, figure out how to install a shower powered by the sun, make a banner, support strikers, pull a prank to make someone laugh, as well as think.

This book is a call to get involved in practical action and reflection to create more sustainable and fairer ways of living. Part handbook, part critique, it is designed to inform, inspire and enable people – you, the person sitting next to you on the train, your neighbour, your mother, your children – to take part in a growing movement for social change. It is us that can make the changes and it is us that will have to.

We believe that this social change is best understood through experiences and real human stories, not abstract ideas. Nine different themes are explored in this book where people are struggling to wrestle back control and build more equitable and just societies – *sustainable living, decision making, health, education, food, cultural activism, free spaces, media* and *direct action.*

This is not a book about a grand unfolding of a new theory on social change or a way to sign up to membership of a political party or campaign group – 'Give us £10 and we'll save the world for you'. It is not a restatement of what is wrong with the world (there are many fantastic books out there that do that already) or about the need to overthrow governments or take the reigns of political power. It's about what we can all do about the challenges we face in the world and how we can make governments and corporations increasingly irrelevant.

Although there is a sense of urgency about what we are saying, there are huge challenges that stand in the way of empowering people to take control collectively. The process won't necessarily be easy and this book does not intend to glamorise what the editors and contributors know can be very hard work. While the book talks about the urgent need for change there are a number of tensions to deal with in making these ideas more accessible and less intimidating to people. There are also many competing voices and visions in the struggle for a better world. Stating the case for managing our own world collectively is difficult as many people ask why they should get involved when there are paid politicians to do so. Not only have people deferred responsibility to leaders and bosses but they are distracted by getting on with their lives, by consumerism, celebrities and the humdrum of daily life.

In response, this book has no easy answers but starts from a premise that there is a growing awareness that change is needed and that the way to make it relevant is by mixing resistance and creativity into a powerful movement that is part of everyday life. As mass protests against the current economic system have ricocheted around the world from Seattle to Cancun, beyond the spectacle of the banners, tear gas and riots, when the streets become silent again, ordinary people are doing extraordinary things, learning by doing, imagining and building the blocks of other possible worlds. We can resist the world we live in while at the same time creating the world we want to see. These small acts are the bedrock for real social transformation – as the phrase goes 'be the change you want to see'. They are the starting points for bringing us together to build our lives outside the logic of capitalism.

if you're not pissed off, you're not paying attention!

Have you ever had the feeling that something is very wrong with the world? Maybe it is the escalating war in the Middle East, or the war on terror (what many call the war of error), people and communities divided by fear and mistrust of each other, that 20 per cent of the world's population use 80 per cent of the world's resources, the potential of catastrophic climate change. Perhaps it is bullying bosses, working long hours for poor wages, while multinationals continue to bleed people and their lands for profit. Maybe it's because public services are being privatised and politicians don't listen, penthouse lofts replace public spaces and house prices soar whilst the majority live in a constant spiral of debt. Or that cases of cancer and stress related diseases, mental illness and depression continue to grow. How have we got to this point?

In twenty-first-century democracies, the way that we are supposed to change something or do 'anything' is to vote – our lifetime's supply of democracy is ten crosses on a ballot paper. Don't worry, commentators say, our political representatives will make the changes we want for us, all we need to do is write to our elected representative and ask them. Yet politicians cannot solve the problem because they are an intrinsic part of the problem, influenced by big business, the possibility of advancing their careers and becoming directors of multinationals, they are at the heart of rotten political systems that largely serve free market policies of neoliberalism.

It is no wonder then that voting has declined and cynicism and apathy have grown as politicians move further away from the everyday needs and desires of people. Politicians seem out of our control as they respond to multinational corporations or institutions like the World Trade Organisation, International Monetary Fund and World Bank whilst unfair trade, poverty and pollution continue unabated.

The voices that dare to say that it is the economic system which is the underlying reason for inequality, climate change and environmental degradation are often disregarded as extremists. The large lobbying non-governmental organisations (NGOs) and charities are able to tell us about the problems, but are unable or unwilling to challenge the root causes of these problems for fear of losing memberships and consequentially their funding. As Robert Newman asserted in the *Guardian* (2 February 2006): 'Many career environmentalists fear that an anti-capitalist position is what is alienating the mainstream from their irresistible arguments. But is it not more likely that people are stunned into inaction by the bizarre discrepancy between how extreme the crisis described and how insipid the solutions proposed?'

One of the biggest crises that we face is the loss of the commons – the common assets that we have built up through centuries of struggle. All of us are facing what is called a 'new enclosure' where our communities and lives are enclosed and privatised as they are stripped of the items essential for life – be it clean water, access to decent education, affordable housing, or enough land to grow food. The problems of a global economy based on fossil fuelled infinite growth can seem overwhelming. But its inherent unsustainability can give energy for creating new ways of looking at the world and organising society. The questions, ideas and potential solutions that the contributors of this book pose explore these possibilities.

everyday revolutions

> And we live in unnatural times.
> And we must make them
> Natural again
> With our singing
> And our intelligent rage.
> (Nobel Laureate Ben Okri, for Ken Saro-Wiwa, 1995)

There are many different words and traditions that could be used to describe where this book is coming from. Our focus concerns 'doing it ourselves'. It is about a revolution that takes place everyday amongst all of us rather than some huge event led by a small vanguard in a hoped-for future. Not waiting for bosses, politicians or experts to take the initiative but building at the grassroots – empowering ourselves and improving our own realities – not to become individual entrepreneurs or free-marketeers, but to work together to make open, sustainable and equal societies.

The principles of the book largely follow anarchist/autonomist thought. *Anarchism*, from the Greek 'without government', is a belief that people can organise society for themselves without formalised government. It argues that the best way to organise is through voluntary arrangements where people are likely to co-operate more. The word *autonomy* is from the Greek 'to self-legislate'. It originates from a strong European tradition, especially Italian and German, and more recently Argentinian, where experiments of how people organise their own lives are widespread.

Another influence of the book is the 'movement of movements', the term that is applied to the loose network of hundreds of social movements, groups and thousands of people who are part of what is widely known as the movement for global social and environmental justice. This book draws on the links, projects, influences and connections between groups throughout the world who make up these movements. The connections are unpredictable and fast changing, but they allow a huge amount of ideas and experiences to be exchanged around the globe. These movements are diverse and are not looking to build global federations or leaderships. That would repeat the mistakes of the past and the present. But there are some common principles which most would agree with: a rejection of borders and nation states, along with wars, exploitation and injustice; reducing over-consumption and the imbalance in the distribution of the world's resources; working towards societies that uphold the dignity of all and the inclusion of everybody; and the promotion of equality and action in everyday life to take back control.

Why do people get involved in such ideas? Often what motivates us are emotional responses – anger, fear, passion, desperation and hope. We all have a right to be angry at injustice, at oppression. Building movements and groups of change is about using this anger constructively. Not falling into traps of hate, powerlessness, blame and desperation but turning those emotions into ones of defiance and strength, hope and inspiration and to intelligent rage.

This book is an advocate for fundamental change, not for seizing power but challenging the way power operates and is linked to wealth and private power in our society. The past has shown us that often seizing power has meant replicating the very systems of oppression that revolutionary movements have struggled to overthrow. In creating fundamental change we need to use a range of methods and tactics and in this our imaginations are our only limitations. There's no simple cause and effect – thinking that if I do this, then that will happen. Many people carry a heavy burden of expectation, waiting for the 'big day' or the 'revolution' after which everything will be all right. The reality is much slower and unpredictable, there's no straight path to where we want to go. Change means constantly evolving, questioning and exploring.

On an everyday level there will always be contradictions and compromises. We might feel alienated by consumerism and work but we seem to have little choice but still be a part of them. Often our lives straddle the real world where we live and work and the ones we dream about and see occasional glimpses of. Of course sometimes it's necessary that we get formal employment, take a flight, buy corporate goods, compromise our ideals. Other times there's scope to be more independent, confrontational and defiant. Asking people to choose between these positions is divisive and unhelpful. Doing something is better than doing nothing at all. Collective organising will not get rid of inequalities overnight or change some of the most destructive things about this world but there are concrete steps that we can all take on the path to moving towards our hopes and visions and to claiming some control over our lives. The experiences reflected in this book show that people are chipping away at the present and building new worlds from inside and outside. The ways of organising our lives, which we look at in this book, can help us realise our potential, create new bonds and offer new answers, and it is there that the potential lies.

competing voices for change

Brian: Excuse me. Are you the Judean People's Front?
Reg: Fuck off! We're the People's Front of Judea.
(Monty Python's *Life of Brian*, 1979)

The arena of struggle for social change is jam packed with competing groups – all with different versions of how the world could be, and how we could get there. A little like the caricatures of Monty Python's *Life of Brian*, there are divisions, suspicion and distrust between those on the left. There is often little that unifies them with groups only coming together when there is a common enemy such as a meeting of the G8 or anti-war protests. It is certainly true that socialists, communists, Trotskyists, anarchists, ecologists or libertarians all differ in terms of tactics and ideas as to where power lies, who is to blame, what is to be done and how it can be done – creating a massive range of responses and groups. The Social Forum meetings and the autonomous decentralised events occurring alongside, the People's Global Action meetings and the many other global gatherings and seminars that take place regularly, have become places for a multitude of voices and responses to meet and exchange ideas.

One of the biggest divisions is between what are sometimes called 'verticals' and 'horizontals'. Verticals usually refer to large and centralising socialist party politics, with a clear leadership or vanguard who aim to mimic and ultimately seek government as a vehicle to achieve emancipation for the masses. They seek to incorporate all 'left' protest movements into their party political model and act as 'the' voice of resistance. Verticals see horizontals as naïve and poorly organised and have often sought to take over their campaigns and take glory for their successes.

Horizontals suggest embedding change in everyday life by rejecting leaders, hierarchy, authority, centralisation and manifestos. Within horizontal politics there is no real desire to take the reigns of state power for fear of repeating its mistakes and taking on its violent tendencies. However, rejecting leaders and clear organisational structures can mean that groups become scattered, and virtually hidden from the public at large. One of the easy criticisms aimed at horizontals is that without leaders and significant organisational structures little can be achieved to really tackle power imbalances. The public and the media often look for them and when found lacking cry that horizontal movements are disorganised and powerless. But this type of coherence is not what many groups aspire to. They prioritise the process – real or direct democracy – more than the end. It is a process which is always in the making. It does and should look messy and unfinished. This rough and readiness is a central part of the politics of horizontality. It is a choice that nobody can represent you. Behind these caricatures, there are actually many shades of grey on both sides, with a history of groups often pragmatically working together on campaigns, publications and meetings.

Those also committed to self-organising must respond to models of left-wing leadership which do offer tangible results. Latin America has a history of popular leaders from Cuba, Nicaragua, Venezuela, Chile and now Bolivia where traditional people power, left-wing movements and military might have brought significant improvements to ordinary people. They have become an exciting cause célèbre for those looking for answers due to their stand against neoliberal policies and the geopolitical ambitions of the United States. In our excitement we shouldn't embrace them uncritically but see what lessons they have for us to manage our own lives. In the last instance we can't rely on military men for our freedoms unless they are prepared to hand over real power to ordinary people.

Additionally, the messages from both established left groups and more horizontal libertarian groups face tough competition from more right-wing and populist groups who advocate self-management in an attempt to take back some kind of control. Both nationalist groups like the British National Party (BNP) and fundamentalist religious groups (both Islamic and Christian) spring to mind. This book does not advocate for such groups. It is important to reclaim a vocabulary of self-management and autonomy from organisations that look inward and seek to control people, lands and resources through violence, dogma and fixed notions of who is in and out, or good and evil, rather than allowing people to govern themselves in free, open societies.

inspirations and struggles

These ideas haven't been plucked out of the air but are part of vibrant, interconnected and often contradictory movements based on rich veins of thought including Marxism, anarchism, syndicalism, socialism, Zapatismo, ecology and anti-capitalism to name a few. History gives us many inspirations: the Diggers, who established a land-based community during the English civil war, the Paris Commune in the French wars of 1871, intentional land-based communities formed in response to the excesses of industrialism, self-organised militias during the Spanish civil war, students on the barricades in May 1968. Latin America has always stood as an inspiration for those struggling against oppression, and in its attempts to stand up against US geopolitics from Allende in Chile and the Sandinistas in Nicaragua. The *piquetero* and unemployed movements in Argentina and the Zapatista autonomous municipalities in the Chiapas state of Mexico are some of the most inspirational examples of how resistance and creativity in Latin America is developing.

There are other more contemporary examples across Europe, such as squatting movements, housing co-operatives, rural and urban sustainable living projects, and radical art and social centres, info shops and bookstores, traveller and free party scenes. We have also seen a growth of strands of 'anti-capitalists', 'anti-globalisers' and the 'global social justice movement' which have now entered into the language of the mainstream media. These clumsy labels at least refer to a growing awareness that people are opposing a broader system of inequality which has a long history. Horizontal networks such as People's Global Action, Reclaim the Streets, Earth First! Permaculture networks, Free Schools, and No Borders groups have spread across the world, campaigning and taking action for radical social change that is based on freedom, co-operation, justice and solidarity and against environmental degradation, neoliberal exploitation, racism, homophobia and patriarchy. New forms of media and the internet have permitted these struggles to become global and the Indymedia concept has spread around the world. Francis Fukuyama's proclaimed 'end of history' thesis declaring the triumph of market capitalism after the fall of the communist bloc has been challenged by these protests.

The Trapese Collective formed in the run-up to the mobilisation against the 2005 Group of 8 summit, as part of the Dissent! network. We undertook popular education workshops to engage with a wide range of people about the problems highlighted by the G8, and more importantly the workable alternatives which we felt were hidden from view. We had been involved in self-managed social centres, intentional communities aiming for sustainability, solidarity with groups like the Zapatistas, direct action against roads, the anti-war movement, climate change campaigning, Social Forums, independent media projects and community gardens. The aim of our work is to engage people in a debate where we all question our assumptions and look at how we can organise and respond. We see this book as part of the laying down of collective achievements, histories and inspirations of autonomous, horizontal politics, and a reflection of how to move forward.

what's in the book?

The 18 chapters that follow weave together analysis, personal stories and examples of various everyday movements for change. The idea is not to dictate how things should be done but to provide examples of places, ideas, ways of organising and inventions for you to do it yourself. This is not a comprehensive guide. Issues such as transport, alternative economies and housing, for example, could have been covered but due to space constraints are not. Neither was there room to do justice to the complex ways that race, gender, class, spirituality, religion or sexuality interact with what we are saying. Although as individuals we have been influenced from many places and peoples, this book is unashamedly focused on what we have found in parts of the world where we have lived. To analyse the global South, especially Africa and Asia, is a mammoth task and was outside our capabilities and experiences. What we have done instead is to focus on where we are best connected and knowledgeable, and leave it to others to do similar kinds of work elsewhere. We wanted to show that there are inspirational examples on everyone's doorstep and it's not necessary to travel the world to find them.

For each theme of the book there are two distinct chapters: one which introduces the theme, its histories, ideas, problems and pitfalls, and a more practical 'how to' guide consisting of ways to turn some of these ideas into reality. These guides are not the last word on the subject, but represent a commitment from people to tell it as they have lived it. Depending on your personal experiences, some things may seem obvious, others less so. We hope you try them collectively, use them to put your version of the future into practice, connect with other people who are doing similar things, or come up with adaptations or improvements. Sources for further reading and research are given at the end of each theme. It is impossible to measure how these ideas may penetrate and travel. For some people they will ignite a spark, others may dismiss them. In any case, the book is a call to action and reflection. It outlines interventions which grow more urgent day by day in face of the crises that loom ahead. A handbook isn't enough to take back control but it's a starting point to get involved with people, networks and movements of resistance, inspiring creativity and changing our world.

resources

Carter, J. and D. Morland (2004). *Anti-capitalist Britain*. London: New Clarion.

Harvie, D., B. Trott and K. Milburn (2005). *Shut them Down!* Leeds: Autonomedia.

Holloway, J. (2002). *Change the World Without Taking Power*. London: Pluto Press.

Kingsnorth, P. (2003). *One No, Many Yeses: A Journey to the Heart of the Global Resistance Movement*. London: Free Press.

Marshall, P. (1991). *Demanding the Impossible. A History of Anarchism*. London: Harper Collins.

Mertes, T. (2004). *A Movement of Movements. A Reader.* London: Verso.

Monbiot, George et al. (2001). *Anti-capitalism: A Guide to the Movement.* London: Bookmarks.

Notes from Nowhere (eds) (2003). *We Are Everywhere: The Irresistible Rise of Global Anti-capitalism*. London: Verso.

Polet, François and CETRI (2004). *Globalising Resistance. The State of Struggle.* London: Pluto Press.

Saad-Filho, Alfredo (2002). *Anti capitalism. A Marxist Introduction*. London and Sterling, VA: Pluto Press.

Schalit, J. (ed.) (2002). *The Anti-capitalism Reader: Anti-market Politics in Theory and Practice, Past, Present and Future*. New York: Akashic Press.

Sen, J., A. Escobar and P. Waterman (2004). *World Social Forum: Challenging Empires.* New Delhi: Viveka Foundation.

Sheehan, S. (2003). *Anarchism*. London: Reaktion Books.

Solnit, D. (2004). *Globalise Liberation. How to Uproot the System and Build a Better World.* San Francisco: City Lights Books.

Solnit, R. (2002). *Hope in the Dark*. London: Verso.

Starr, A. (2004). *Global Revolt*. New York: Zed Books.

Tormey, S. (2004). *Anti-capitalism: A Beginners Guide*. Oxford: One World.

Wall, D. (2005). *Babylon and Beyond*. London: Pluto Press.

Walter, N. (2002). *About Anarchism*. London: Freedom Press.

Yuen, E. et al. (2005). *Confronting Capitalisms*. Brooklyn, NY: Soft Skull Press.

1 why we need holistic solutions for a world in crisis

Andy Goldring

The premise of this chapter is that our world is facing massive ecological crises, as well as the potentially disastrous social and economic problems that stem from this. In understanding how we can change our world it is important to outline some of the enormous problems it faces and every species that inhabits it. The point of this chapter is not to feel overwhelmed by the extent of the problems, but to examine existing, easy to implement and inspiring approaches that we can use to both improve the environment and the lives we lead, looking at the holistic approach of permaculture in particular as a mechanism in creating change. Sustainable living is more than just a nice life for those that attempt it. It also offers a vision of a better world, and a daily, practical protest against the cultural, corporate and state structures that lay waste to the world.

the ecological crisis and how we got here

That we are living within a rapidly escalating ecological, social, political and economic crisis is beyond doubt. This has been outlined rigorously over the last few decades in reports like *The Limits to Growth* (1972), *Our Common Future* (better known as the *Brundtland Report*) (1987) or landmark books like Fritz Schumacher's *Small is Beautiful* (1973), as well as the WorldWatch Institute's annual *State of the World* report (www. worldwatch.org). How we came to be in this situation is less certain. A summary of my view, informed by 15 years of practice working with the Permaculture Association is this. From its earliest beginnings, humankind lived in relative harmony with nature, ruled by its laws, in tune with the seasons and with minimal disruption to the overall ecological system. At the end of the last ice age, climatic conditions changed and productivity increased, and humans in the Middle East, East Africa and China moved from gathering and hunting in small groups to settled agriculture. Impacts were huge and many writers of social and human ecology such as Murray Bookchin, John Zerzan and Michael Sahlins see this as the origins of our present civilisation and its trappings such as hierarchy, division of labour, oppression, trading and specialisation, more complex social organisation, and the first cities. Ultimately these civilisations were unable to manage their resource base and failed. Reasons included soil and tree loss, the collapse of agriculture, war with competing civilisations or an inability to change inappropriate social and environmental practices. These ideas have been eloquently outlined by Jared Diamond in a number of books such as *Guns, Germs and Steel* (1997) and *Collapse* (2005).

The 'Medieval Warm Period', alongside new inventions from China, such as the horse chest harness, in the tenth to fourteenth centuries enabled increases in European agricultural yields and the rapid expansion of larger human settlements. Other Chinese inventions such as gunpowder, paper, printing and the compass also had a transformative effect on medieval society. The combination of increased agricultural yields and new inventions enabled small European kingdoms to form the first nation states. Environmental and social limits were overcome through colonial expansion into new lands. The use of millions of mainly African slaves during the seventeenth century allowed companies to exploit the new lands and set up vast trading empires. During the seventeenth and eighteenth centuries inventor-scientists started to harness the power of water in new ways, with a major leap in industrial capacity occurring when the power of coal was harnessed to create steam engines. The Industrial Revolution had begun, and ushered in a new scale of environmental and social change. Companies flourished and became huge enterprises. The 'enlightenment' and other philosophical movements decided that humans were above nature and therefore it was ours to exploit as we saw fit. In the nineteenth century, we discovered a seemingly limitless supply of easily transportable explosive energy in the form of oil. Human population levels soared and a considerable middle class emerged with aspirations for comfort and a huge appetite for consumer goods (see classic works such as E.P. Thompson's *The Making of the English Working Class* (1968). World wars, the worldwide industrialisation of farming through the petrochemical-based 'green revolution', the 'triumph of capitalism' across the globe backed by new forms of international financial institutions like the World Bank and IMF, and mass media-based propaganda completed our divorce from nature and left most humans reeling from the effects of over-consumption or a life of poverty. There are many excellent commentaries which outline these changes and are included in the resources listed in Chapter 2.

At the beginning of the twenty-first century we face a huge list of interconnected challenges. Here are just a few of them:

★ Climate change: The burning of coal, oil and gas, and the clearance of forest for agriculture is changing the climate through the 'greenhouse effect' and may soon reach a 'tipping point' beyond which humans can have no influence. There is now widespread agreement that climate change is the most urgent challenge facing the planet. The Stern report of 2006, written by former World Bank Chief Economist Nicholas Stern, suggests that there is now a 50 per cent chance of temperatures increasing by 5 °C, with catastrophic consequences for every species on the planet.

★ Peak oil: A term popularised by scientist M. King Hubbert who, while working for the US Geological Survey, suggested that the world's supply of available oil would peak between 1990 and 2000. He got the date slightly wrong, but there is now wide consensus that we are within a few years of 'Hubbert's peak', with the gas peak following 15–20 years behind (Heinberg 2005). As a result of this peak the energy foundations of industrial society are dwindling. 'Alternatives like biofuels, ethanol or biomass can play a marginal supportive role but nowhere near on the scale required. When the oil runs out the economic and social dislocation will be unprecedented' (Michael Meacher, former UK Environment Minister, quoted in www.peakoil.net).

★ Water: Water shortages and drought are becoming more prevalent, with many ancient aquifers that take thousands of years to recharge, near full depletion. 'Global freshwater use tripled during the second half of the twentieth century as population more than doubled and as technological advances let farmers and other water users pump groundwater from greater depths and harness river water with more and larger dams. As global demand soars, pressures on the world's water resources are straining aquatic systems worldwide. Rivers are running dry, lakes are disappearing, and water tables are dropping' (Elizabeth Mygatt, 26 July 2006, 'World's water resources face mounting pressure', Earth Policy Institute website). It is clear that water will be a key resource and a source of war and conflict in years to come.

★ Industrial agriculture: Industrial farming has caused the destruction of whole ecosystems, made many species extinct and laid ruin once highly productive agricultural land. One recent example is our new demand for biofuel, which is leading to the widespread destruction of Indonesian rainforest and peatland to make way for huge monocultures of oil palm, all under the guise of an 'eco-solution'.

★ Ocean ecosystems: Chemical agriculture, unprocessed sewerage and industrial fishing have created 'dead zones' covering hundreds of thousands of square kilometres of ocean. The majority of fish stocks are in decline. Global fish stocks could be almost eliminated within 50 years if current trends continue.

★ Soils: Soils have been depleted of minerals leading to poor food quality, increased disease and fire susceptibility. Poor soils and diminishing water supplies are contributing to famines that now ravage many regions and are set to worsen with climate change.

★ Environmental refugees: The number of environmental refugees is mushrooming as people desperately try to escape from areas no longer able to meet their basic human needs. The New Economics Foundation estimate that by 2050 there will be over 150 million environmental refugees unless pre-emptive action is taken. There are many underlying causes to this refugee crisis which go beyond short-term droughts: people are forced from their lands by wars often fuelled and funded by ex-colonial masters in the West, the expansion of cash crops continues to deprive people of land and force movements, and the effects of climate change such as long-term drought, flooding or extreme weather events is increasing mass movements of people.

★ Ownership: Corporate control of key resources and utilities, such as seeds and water, undermine local efforts for self-reliance, and fuels the growing gap between rich and poor, both nationally and globally. The profits of Royal Dutch Shell now equal the GDP of Egypt (*Guardian*, 7 November 2006). George Monbiot in *Captive State* (2000), John Pilger in *The New Rulers of the World* (2002) and Naomi Klein in *No Logo* outlined the powerful role corporations play in shaping our lives.

★ Culture and society: Our societies are largely based on a dysfunctional cultural model which is difficult to comprehend as it is so all-encompassing. Some of its premises include: endless economic growth, the primacy of profit and growth and the need for production and hence consumption, global trade and a wage economy which fosters individualism and competition, the nuclear family and all its trappings of suburban and out of town developments, and an education system that largely trains people for participation in narrow work tasks. This leaves us largely divorced from nature and each other, at the whim of corporations, neglectful states and global institutions, and a media system in the hands of dull, powerful companies that obscures rather than illuminates.

Struggling for sustainability

Given the enormity of the problems that are faced, many people are going to ask the question 'Is there any hope?' Is it possible to fundamentally change the economic/

industrial/military system? Can we move from a society based on
the pursuit of power, profit and consumption to a society that has
the well-being of society and the environment at its core? Can
this be done at a global level? Is it fair to curb the Western style
'development' in other parts of the world, especially Africa and
Asia? These are difficult questions to answer, but in my opinion, yes there is hope.
All the ideas, techniques, technologies and cultural models we need to transform the
world and steward the environment for the better exist already. They have developed
throughout history and can be seen through several currents.

Firstly, there are the sustainable practices of human scale societies. These groups,
generally numbering fewer than 300, meet the majority of their needs from within
their own region. Human scale societies – both nomadic tribes and settled villages
– were more prevalent in the pre-industrial world and made up the majority of the
human population until just a few hundred years ago. They worked less than we do,
met their needs without destroying their environment and had no need for standing
armies or police forces. Their whole way of life was tuned to the local environment,
each generation, from children to elders, had a role to play and everyone contributed
to the well-being of the whole group. Strong group identity, strict taboos and an
appreciation of the 'web of life' ensured that their way of life was sustained over
hundreds of generations. However, we must also be careful not to over-romanticise
such human scale societies as some kind of ideal template – no doubt they faced a
different set of problems, such as food shortages, occasional outbursts of bloody
conflict, more reliance on manual labour, none of the comforts that mark out
consumer-based societies, nor the ease of mobility that we enjoy.

Having said that, there is much which our profoundly industrialised societies can
learn about regaining a sense of simplicity, social integration, and cultural approaches
to living within natural limits, autonomy and self-reliance. In her book *Ancient Futures*
(1992) Helena Norberg-Hodge outlines how a planned process of development
since the 1970s based around military expansion, tourism and resource use has
undermined such human scale communities in Ladakh, northern India. Here, the
Ladakh Project has been set up with the aim of what it calls 'counter-development' to
re-establish the viability of these human scale communities connected to the rhythms
of nature and regional trade and agriculture. Strong tendencies towards these types
of human scale communities clearly still exist into the present, seen through this
example and moves towards ecovillages, co-housing developments and intentional
communities. The interesting question that arises, and what this and other chapters
address, is how do we create the conditions to make these kinds of communities more
viable on a wider level?

Secondly, there are groups that have rebelled against the ideas and power structures of the time, or developed new ideas that are then adopted by society at large. Even thousands of years ago, Socrates observed environmental destruction and called for the widespread reforestation of Greece. In the seventeenth century, the Diggers struggled to create a more democratic and fair society, and show that freedom from poverty, hunger and oppression could be won if the earth were made a 'Common Treasury for all'. In the nineteenth century, the Luddites rebelled against the new machines of the Industrial Revolution but were quickly quashed by the state and the 'march of progress' (see Christopher Hill's *The World Turned Upside Down* (1972) or Kirkpatrick Sale's *Rebels Against the Future* (1995)). By the twentieth century, the problems had become bigger, but so had the movements that sought a better way. The science of ecology led to a new appreciation of nature, organic agriculture re-emerged, natural farming was pioneered by people like Masanobu Fukuoka and Wes Jackson, and the self-sufficiency/back to the land movement was championed by John Seymour and others. Rachel Carson and her seminal work *Silent Spring* (1963) provoked a new interest in caring for the Earth and an ecological movement based around membership groups like the Sierra Club, Greenpeace and Friends of the Earth, as well as direct action groups such as Earth First! emerged.

permaculture as an holistic solution

> Permaculture is revolution disguised as organic gardening. (Mike Feingold, community activist and designer)

Whilst many techniques and technologies for solving specific parts of our multifaceted problem exist, there are very few integrated or 'holistic' approaches that aim to tackle the problem as a whole. One such integrated approach is permaculture. The term was coined by two Australians, Bill Mollison and David Holmgren, to describe an ecological design approach to sustainability, and has been spreading across the world since the late 1970s:

> Permaculture is the conscious design and maintenance of agriculturally productive ecosystems which have the diversity, stability, and resilience of natural ecosystems. It is the harmonious integration of landscape and people providing their food, energy, shelter, and other material and non-material needs in a sustainable way. (Bill Mollison 1997, ix)

Permaculture has three main ingredients:

1. Ethics

   ☆ People care: People care is about looking after yourself and the people around you and ensuring that your actions don't harm other people you don't see, such as when you buy food produced by workers on low wages using health damaging chemicals. It is also about considering our legacy and working to make the world better for future generations.

   ☆ Earth care: Opposition to further ecosystem destruction, rehabilitation of damaged land and a commitment to meet our needs on the smallest amount of land possible, so that we can leave space for all other species.

   ☆ Fair shares: This stresses the redistribution of skills, resources and money to enable more earth care and people care. It is also about limiting our consumption to that which the earth can sustain.

2. Ecological and attitudinal principles
   Key principles include: direct observation of natural systems and an increased understanding of how they work; relative location because creating beneficial functional relationships between different elements within a system is vital; the support of important functions by many elements to ensure diversity and resilience; the provision of many functions by each element (for example, a shed becomes a water harvesting surface).

3. Design

   Permaculture provides a new design language for observation and action that empowers people to co-design homes, neighborhoods, and communities full of truly abundant food, energy, habitat, water, income ... and yields enough to share. (Keith Johnson, editor/writer *Permaculture Activist*).

   Design is where we put our ethics and guiding principles into practice. Design is a 'pattern', a 'plan of action' that enables you to make better use of your existing resources through improved placement and new relationships, and helps you develop new ways to meet your needs. The great thing about permaculture is that you can start wherever you are, whatever your situation. If you are in a high-rise block, or a 1000 hectare farm, you can design the environment around you to become more sustainable and productive, and less polluting. You could start, for example, by growing food on your windowsill.

   The Permaculture Flower in Figure 1.1 highlights these features.

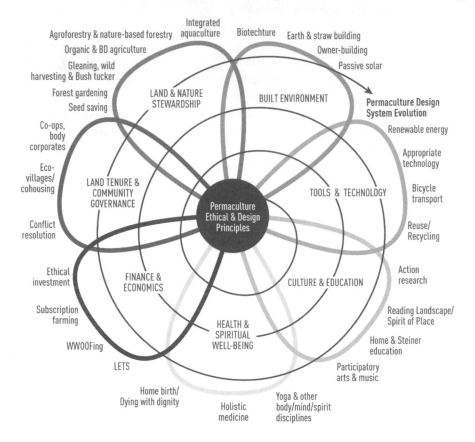

Figure 1.1   The permaculture flower

Source: David Holmgren.

Characteristics of permaculture systems

Whilst every permaculture design is unique – tailored to the specific landform, climate, and requirements of the site and its inhabitants – all designs share some common characteristics:

★ Localisation. As oil and gas become less abundant and more expensive, we will need to meet our needs locally. When apples are flown from New Zealand to the UK they create up to 120 times more pollution than apples grown in the UK. So whether it is because we want to stop climate change or prepare for peak oil, create local livelihoods or just enjoy the fruits of a productive healthy landscape, we need to localise the systems that meet our needs.

★ Self-reliance. Permaculture is not about self-sufficiency. I can't knit jumpers, but I'm happy to exchange the things I'm good at with others. Do what you can, co-operate with others and aim for a largely self-sufficient region.

★ Decreasing the need for external inputs of energy and resources. Through the application of permaculture design and the development of local self-reliance, we reduce the amount that needs to be imported to meet our basic necds. By farming without chemicals, recycling resources locally, changing our eating habits, and celebrating and supporting local creativity we move from dependency to interdependency.

★ Use of renewable energy systems. Our first aim should be to reduce the amount of energy that we need to meet our basic needs. When this has been done we can move on to develop local energy systems to meet those modest needs. Long-term forestry, passive solar construction and solar water heating are priorities, with optimum electricity generation dependent on local circumstances.

★ Design as an active and ongoing participatory process. It will take a long time to create a sustainable human society, and quite a long time to get our own house in order. There is always an opportunity to improve what we do, and support others to improve what they do. I know of nowhere that is finished and fully sustainable.

sustainability in action

So how do these ideas work out in reality? There are countless examples of permaculture in practice, from the tiny to the town-wide. Long-term 'Energy Descent Action Plans' are springing up in response to crises to manage the energy descent after the peaking of oil. Based on a model developed in Kinsale, Ireland, plans are now underway in the UK in Totnes, Devon and Lewes, Sussex (see www.transitionculture.org). Four case studies follow, from the everyday to emergency situations, which show how sustainability is being put into action simply and effectively in a range of settings to tackle key issues like energy, water, food and waste.

The Yellow House, UK. 'Turning our house into an low energy eco-home'
George Marshall, environmental campaigner and member of the Climate Outreach
Information Network, reflects on turning his humble house in Oxford, UK, into an
eco-home.

When Annie and I moved into our new home, 1930s ex-council terraced house in
Oxford, it needed a lot of work. Our dream was to renovate the house so that it is
energy efficient and respects the environment, but is also clean, healthy and full of
natural light. We followed the usual advice – new insulation throughout, lagging
hot water pipes, draughtproofing and a new condensing boiler. We found it cost
very little to far exceed the recommended levels of insulation. In the new extension
and loft conversion we doubled the level of insulation required by building controls
for the cost of a few hundred pounds/euros/dollars more in materials.

We managed to save money by using salvaged materials everywhere – for joists,
floors, sinks, stairs, and light fittings. We built the fitted kitchen from old furniture
– a beautifully made solid oak kitchen for less cost than the cheapest chipboard
kind. We installed many eco-features. We have a high quality solar hot water
panel. The extension has a grass roof and a sun porch to preheat the air entering
the house. The bath water is stored in a copper tank (a reuse of the old hot water
tank) and runs the downstairs toilet. The upstairs bedroom operates as a sun trap,
heating air which is then pumped downstairs by a thermostatically controlled fan.
And all paints, floorings, and timber meet the highest environmental standards.

So we managed to add two new rooms to the house and still reduce gas, electricity
and water consumption by nearly 60 per cent. This is the UK Government's target
for 2050 and we achieved it within one year! It is clear that soon every house
will have to do what we have done. But a house is only one part of one's overall
emissions. We try to achieve a low carbon lifestyle in other respects. We eat only
local free range meat and local organic vegetables, rarely buy something new,
work within cycling distance of home and only use a car for rare and unavoidable
purposes. Providing we can avoid flying, which is a massive way to cut your carbon
use, we manage to keep our annual emissions under 2 tonnes of carbon dioxide
per person, which is not far off the sustainable level. You can learn more about low
carbon lifestyles and work out your emissions on the calculator at www.coinet.
org.uk/motivation/challenge/measure.

The most important thing we have learned is that our house can be a model
for inspiring others. We had so much interest that I wrote a website so that people
could make a virtual tour of the house and learn from our experience and contacts.
The site has had over 10,000 hits and we get visitors and letters every week. We

have also produced a CD-ROM which can be ordered through the site (www.theyellowhouse.org.uk).

The HoriZone at the 2005 G8 summit: shit and social change
Starhawk, writer, activist, trainer and permaculture designer, writes about her experience working at the HoriZone protest camp when 3000 people gathered to protest during the G8 summit in Scotland 2005. There she developed regenerative design through compost toilets and grey water systems.

During a recent project at a large encampment, I was explaining one aspect of compost toilet maintenance to two young women. These particular toilets were basically a framework set up over wheelie-bins. The waste drops into the bins, sawdust is added after each deposit, and when it's full, the bins can be wheeled off, sealed, stored for two years, and the resulting compost is then safe to use on landscape plants and trees. It's one of many possible ways of dealing with human wastes.

'Often the "deposits" pile up in a sort of peak in the middle,' I said. "But if you reach in with a stick and push it around, you can get it to fill more evenly.'

The first young woman looked at me with a kind of horror in her eyes. "We can't even get the blokes to do the dishes," she said. "How are we going to get them to stir shit?"

I could have said, but didn't, that stirring shit up, on every level, is the basic work of anyone wanting social change.

'But it does make you think,' the second woman said. 'How privileged we are in our ordinary lives ... that we never have to think about where our waste goes or where our water goes.'

Watching that realization dawn in her eyes, I knew all the work we had done to make the camp happen was worth it.

The HoriZone ecovillage was an attempt to demonstrate, first to ourselves that we can deal with the real shit: providing for basic human needs in a way that respects and even regenerates the environment. Our directly democratic structure used at the camp was a way to also show that real work can be organized without bosses or hierarchy. Everyone participates in all aspects of the work. I carried loads of wood and drove a few screws into the compost toilets. Collective work for a big camp counters many of our societal myths about work: that hard work is demeaning, that people won't do it without pay, that work is something to be avoided, that unless people are strictly controlled, they'll be lazy shirkers. Work is a way to connect with other people, to feel part of something, and to gain respect and recognition. Most importantly, it shows that we have the skills to manage our own lives and implement sustainable living right here and now.

Earthaven ecovillage, USA

This extract is taken from the Earthaven website (www.earthaven.org), one of the most inspiring ecovillage projects currently up and running, based in south-east USA.

Ecovillages are human-scale, full-featured settlements in which human activities are harmlessly integrated into the natural world in a way that is supportive of healthy human development, and which can be successfully continued into the indefinite future. (Robert and Diane Gilman, *Ecovillages and Sustainable Communities* 1991)

Imagine it's summer at Earthaven, an aspiring ecovillage settlement nestled in the forested slopes of the Southern Appalachians in the USA. Along with murmuring streams and birdsong, you hear the sounds of human activity, of people building their common future together, of children at play... You hear the sound of power tools and home construction, often with lumber from trees felled on the land. This is the sound of liberation. Using our own lumber and hiring each other to build our homes frees us from banks and the timber industry while keeping materials and money within our village economy. These are radical acts. We're learning to practice ecologically responsible forestry and agriculture; to develop natural building systems that sustain forest health, create jobs, and generate renewable energy through good design. We intend to become empowered, responsible, ecologically literate citizens, modeling bioregionally appropriate culture for our time and place.

Founded in 1994, Earthaven is located on 320 acres in culturally rich, biologically diverse western North Carolina, about 40 minutes southeast of Asheville. We are dedicated to caring for people and the Earth by learning and demonstrating a holistic, sustainable culture. Lying between 2000 and 2600 feet elevation, our forested mountain land consists of three converging valleys with abundant streams and springs, flood plains, bottom land, and steeper ridge slopes. We intend to become a village of at least 150 people on 56 homesites. As of 2004 Earthaven has 60 members, with 50 living on the land, including several young families with children. Our permaculture site plan includes residential neighborhoods and compact business sites, as well as areas suitable for orchards, market gardens, and wetlands.

Much of Earthaven is still under construction. Physical infrastructure so far includes roads, footpaths, bridges, campgrounds, ponds, constructed wetlands, the first phase of our water system, off-grid power systems, gardens, our Council Hall, a kitchen-dining room, many small dwellings, and several homes. We govern ourselves with a consensus decision-making process and a Council and committee structure. We own title to our land, which we financed with private loans from members. Members pay their share of the cost of the land by leasing homesites from the community. We value sustainable ecological systems, permaculture design, elegant simplicity, right livelihood, and healthy social relations. We are spiritually diverse. We have both vegetarians and omnivores; some members raise livestock.

Our small ecologically sound businesses include Red Moon Herbs; *Permaculture Activist* and *Communities* magazines; the Trading Post, a general store and Internet cafe; a permaculture plant nursery; carpentry and home construction; tool-rental; solar system installation; plumbing and electrical installation; website design; candle-lanterns and other wooden craft items; and consultants and courses in permaculuture design, natural building, creating new ecovillages, herbal medicine, women's health, and women's mysteries. We teach workshops on starting and designing an ecovillage.

Models for the future: responding to emergency situations

Writer and activist Starhawk explains the importance of developing skills to respond to emergency situations such as those seen during Hurricane Katrina in 2005, which are likely to become more frequent in the face of climate change.

In mid 2005 Hurricane Katrina struck New Orleans, followed by Hurricane Rita which hit the Louisiana coast. The U.S. government's response to these disasters ranged from inept to criminally negligent. For example when the city authorities ordered the evacuation of New Orleans, they didn't provide any transport for those who didn't have private automobiles. In the refuge of last resort, the superdome, where people were told to go, medical staff, supplies, even food and water weren't properly provided. But where the government failed, people's movements stepped in. A group called *Common Ground* arose in the first days after the hurricane, spearheaded by a former Black Panther, Malik Rahim, who lives in the Algiers neighborhood of New Orleans which was spared flooding. They first organized neighborhood protection from vigilante groups which were roaming the streets, then went on to organize garbage pickup, distribution of relief supplies, and a free medical clinic which was the first to be running after the disaster, and which has now served thousands of local people, many of whom had no medical care for decades before the disaster, because they couldn't afford it. The clinic has a warm, friendly and respectful tone, in contrast to those eventually set up by the military and the Federal Emergency Management Agency, which were surrounded by barbed wire and armed guards. Working with Common Ground, some of us are now working on bioremediation projects in New Orleans. We have set up one demonstration project to brew biologically active compost teas that can clean toxic soil, and have launched a training program and larger pilot project in connection with a local community garden. We're raising worms, brewing bioremediation teas, inoculating wood chips with beneficial fungi, and propagating plants that can uptake heavy metals.

If we want to transform the world, we need solid models of how to do it as well as a critique of what's wrong with it. If we want people to move away from the system they know and which has always provided for their needs, they must feel confident that a new system can provide for their security and survival. In New Orleans, where the official systems failed so badly, we have shown that the methods of organizing and the skills we have learned in social movements can indeed provide for those needs, and do so in a joyful, egalitarian, directly democratic way. Natural disasters will increase with global warming. The need for protest and the encampments that go with it will continue as long as injustice continues. But when we can also grasp the opportunities in these situations of disruption for new creation, when we can truly value all parts of ourselves and see even our wastes as a resource, we can tap powerful forces of change and transformation, that can help us regenerate both the human community and the soil, and bring healing and balance back into this world.

facing up to the limits of sustainability

So what's stopping us? If the conceptual frameworks, the eco-technologies and the networks of activists are all there why haven't we turned it all around? At any one time a variety of factors make it hard for us to move forward, and it takes ingenuity to find routes around obstacles.

At a local level, when you're trying to do the right thing:

★ Your local council may be obstructive, through planning constraints, local policies, or hidden agendas.
★ You might feel isolated and unable to make a difference.

☆ You are trapped in a job, tied to the bank by a mortgage and too tired at night to think about anything else.

☆ You can't find land or buildings from which to develop new projects.

The most local and most serious problem occurs when you give up and don't think you can make a difference anymore. But all these limits can be overcome with persistence, clear thinking, talking to others, or turning the computer off, and then on again.

At a global level we come to more troublesome limits, as well as their potentials:

☆ Market forces make local production of food and many of our basic needs 'uneconomic'. However, this will change when oil shortages really kick in and render the globalised movement of goods 'uneconomic'.

☆ Desertification and the transformation of good quality agricultural land to poor marginal land. The development of increased water holding capacity through contour work and reseeding with hardy self-seeding plants and trees offers hope that this limit can be overcome.

☆ The costliness of expensive eco-technologies. However, there is usually a cheaper DIY approach, or a completely different, less technical solution.

☆ Disempowerment and mass hypnosis of the billion or so middle-class inhabitants of planet Earth. This group is destroying the Earth with its affluence, but has the financial resources to transform it. Every T-shirt you wear is an opportunity to get others thinking. Every 'action' and project is an opportunity to get others involved. Turn off the TV and encourage others to do so.

☆ Addiction. New cars, bigger lawnmowers, heroin, cannabis, TV, bigger this, bigger that, a couple of beers every night. Every time we kick an addiction, we notice another. It's good, it means we are evolving and becoming a better person. But to talk about society's addictions, to bear witness to our collective insanity, takes courage and collective action.

☆ Perceived lowering of living standards. Whilst American incomes have grown steadily since the 1950s, happiness has reduced. Remember the mantra, 'quality of life'. It's what everyone wants, what we all deserve, and comes about by travelling a path back to greater connection to nature.

☆ Legislation. Do we stop voting and do our own thing, or engage with others and change the framework within which everyone operates? A very tricky question.

☆ Lack of appropriate skills and education. Our children are still being taught about gravity or why the apple falls off the tree. A more thoughtful and stimulating educational system would ask how it came to be on the tree in the first place. We need to turn state education systems on their heads. Our children really are our future, so we must ensure they are learning the skills they will need to create a sustainable one.

★ Denial. We don't act for various reasons including the messages we receive from corporate funded sceptic-science which suggests everything is under control, being sidelined by greener forms of capitalism, and the psychological strength needed to accept the implications of what ecological crises means for us. None of us can be arrogant enough to think that we have all the answers. What can motivate us is how things can, and do, move quickly when the power of human attention becomes focused.

All these limits can be overcome with persistence, clear thinking, and a new sense of global co-operation.

the future

The current system is collapsing. But in the midst of this chaos, new shoots are emerging, new possibilities trembling and opportunities beckoning. We have seen a range of contexts for these possibilities, from your own house, to larger villages, camps and emergency situations. In my work I come across hundreds of groups and projects determined to do it all differently. They are improving their local environment, meeting a greater proportion of their needs and improving their quality of life. They are also helping to open up new pathways, making it easier for new people to get involved.

The skills of permaculture, natural regeneration, self-organisation and living within limits are easy to learn, simple to start and put into practice. They take persistence, a determination to break old habits, and yes, not everything works out the way we want, but each positive action links us to a new global family that has the interests of the Earth and all its beautiful inhabitants at its heart. We're not alone. Millions of people across the world are working to make things better. When you go to bed after a good day of rabble-rousing and Earth repair, others are just waking up, ready to put in another day's effort. As Hazel Henderson suggested recently, the networks of people that are working for Earth and societal repair, linked by the internet, and a million small agreements to work together, are emerging to form the world's greatest, most important, new global superpower.

It's easy to get involved. Reducing your energy use through conservation measures and maybe even developing your own electricity supply means a reduction in bills, and a reduction in profits to corporate power companies. Car sharing or, even better, giving up your car, means less outgoings, it means cycling, walking and taking more exercise, leading to an improvement in your own health, reduced local pollution and a friendlier neighbourhood. It makes it harder for General Motors to turn a profit. Growing your own organic food gets you outside gardening, puts you in touch with the soil and gives you valuable skills for when oil-based agriculture is no longer viable. If you can't grow your own, support local growers. Cargill and Nestlé can't profit from the lettuce you sow in your window box, nor from seed you saved from the last crop. Localising production of electricity, food, fibres, biomass and the stuff of daily life means improving community well-being and developing a local economy that is better placed to survive the changes ahead. It means you take back control and responsibility for meeting your needs, which connects you powerfully to the rest of the people in your neighbourhood. Self-reliant well connected communities can resist government diktat. It also means producing less pollution and climate disrupting gases.

It doesn't really matter where you start. Follow your curiosity and passion, make it part of your life with practical action and steady learning. Celebrate your achievements and turn others on to the possibilities. There are thousands of organisations and groups to connect with, many simple practical steps you can make right now. There's not a moment to lose, it's more fun than TV, and infinitely better than putting up with business as usual.

Andy Goldring is a permaculture teacher, designer, member of Leeds Permaculture Network and co-ordinator of the Permaculture Association (Britain). Additional material supplied by Starhawk, US based writer, activist and ecologist, George Marshall from the Climate Outreach Information Network and from Earthhaven Ecovillage in the USA.

2 how to get off the grid

Bryce Gilroy-Scott

technology as transformation

Technology is much like the proverbial sword, it can be used to kill but it can also be used to bring life by furrowing the earth to plant a seed. The sword's use depends on the desires and values of its wielder. Appropriate technology is a type of technology aimed at improving the living conditions of those who use it. It is appropriate to communities so that community members can understand how to use, repair and recreate the technology. Appropriate technology is also economically appropriate; the people who depend on it can afford it. It is a form of active resistance to reclaim control of the technologies fundamental to our lives and the raising of our families. These technologies provide our food and shelter and allow us to heat our water and homes. They are essential to our survival.

But consider how you will fulfil your basic food, shelter and heat needs in the face of great changes outlined in the last chapter: cheap oil becoming scarce, migration increasing, climate chaos worsening, pollution and contamination escalating. Survival will become a much greater struggle. The last chapter looked at some of the crises we face and the ideas behind how we can respond. This chapter looks not only at appropriate technologies that can be a significant step towards creating a sustainable society but also at steps we can take in our lives and our homes, today, that move us towards disconnecting ourselves from the centralised grid and providing for ourselves and communities the things that we depend on. The chapter is divided into three sections focusing on energy, waste and water. Other issues such as food are dealt with elsewhere in the book. Reducing the amount of energy or water we use doesn't necessarily mean a reduction in the standard of living – it means using our common sense, consuming responsibly, thinking about our actions and putting back into the earth what we take out. Not everything can be done overnight and for people living in dense cities without access to gardens or land it can be difficult to see what to do. But a lot can be done. For this reason, we include three 'Today' sections, of things you can do right now or in the weeks ahead after reading this book if you haven't already, and then we discuss some more ambitious, creative and fun projects you could take on if you have the space, resources and time. On their own the steps we can take today are not going to tackle the ecological crises we face, but they are important steps towards sustainable living and getting used to the coming energy descent.

Figure 2.1 below shows how some of these modifications, both simple and more challenging, from better insulation to a backyard solar shower, can become part of a typical home.

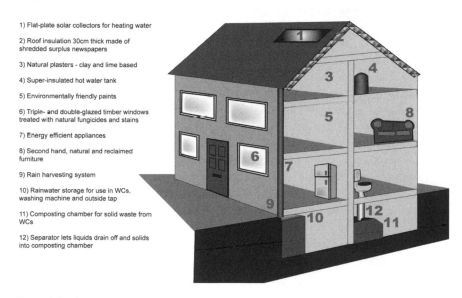

1) Flat-plate solar collectors for heating water

2) Roof insulation 30cm thick made of shredded surplus newspapers

3) Natural plasters - clay and lime based

4) Super-insulated hot water tank

5) Environmentally friendly paints

6) Triple- and double-glazed timber windows treated with natural fungicides and stains

7) Energy efficient appliances

8) Second hand, natural and reclaimed furniture

9) Rain harvesting system

10) Rainwater storage for use in WCs, washing machine and outside tap

11) Composting chamber for solid waste from WCs

12) Separator lets liquids drain off and solids into composting chamber

Figure 2.1 An eco-home

Source: Adapted from Liquoricefish Design.

energy

The majority of us live outrageously energy intensive lives. While many things are necessary, by doing a personal energy audit we can all find things that we can do to cut our energy use, often dramatically. With almost one-third of emissions due to the transport sector this is often where you can save most. This takes will, imagination and initiative but is an essential first step in becoming aware of our own ecological sustainability.

Box 2.1 Today...

Reduce the amount of energy you use: holiday near to home; get the train where possible instead of flying; cycle or use public transport instead of driving; invest in a bike trailer for shopping trips; use car pool or lift share schemes; set up a school walking bus; convert your car engine to biodiesel/vegetable oil; turn off the lights and appliances in any room you're not using; turn equipment off stand by; insulate your loft to 270 mm (12 inches) thick; double glaze your windows and cut heat loss by 50 per cent or get good thick curtains; switch to a green energy supplier; change light bulbs for low energy ones; put on an extra jumper and keep your thermostat constant and as low as comfortable; check for window and door leaks and seal them up; install solar water heaters; super insulate your boiler or get a high-efficiency condensing one; get a wood burning stove to heat space and water; use well fitting lids on saucepans; only boil as much water as you need in the kettle; fit a jacket to your water tank that's at least 75mm (3 inches) thick; insulate walls and cut energy loss by up to 33 per cent; wash laundry at 30 degrees and avoid tumble driers; draught proof skirtingboards; don't buy all that energy intensive stuff you don't really need!

Micro-generation: harnessing nature

Micro-generation is a way of reducing our reliance on natinal energy grids. Many people are experimenting with small, stand-alone systems that harness the power of the wind and the sun. While these won't be able to power energy intensive appliances such as irons and electric kettles for long, creative thinking about what's really necessary means that a household can run most of its needs from a much lower power input. There's not enough space here to go into detail, but here are a few pointers about one example – a small wind turbine.

Firstly, getting your own energy from wind power is attractive as it is free once your system is up and running and with a bit of help and research it is doable. Buying and installing your own wind turbine and batteries might cost anywhere from £1500/$3000 to £15,000/$30,000 depending on your power requirement. Micro hydro systems (small water turbines) are also becoming increasingly popular, they can be designed not to interfere with the flow or biodiversity of the stream and can produce significant amounts of energy. There are many detailed guides available that provide step-by-step instructions (see resource list at the end of this chapter) and a number of places that run courses on the making and installation of turbines. In urban areas wind and water turbines are often not appropriate although planning permission is beginning to be granted for tall buildings with wind turbines on the roof, such as the new London Climate Change Agency Palestra building, as well as many schools and hospitals which are installing turbines on playing fields and carparks. If micro-generation is not possible, there are other possibilities, for example forming an energy co-operative like the Baywind Energy Co-operative.

Box 2.2   Baywind Co-operative Ltd

Baywind Energy Co-operative Ltd was set up in 1996 in Cumbria, UK, to build community wind turbines on the lines of co-operative models successfully pioneered in Scandinavia. It's an Industrial and Provident Society where decisions are made collectively and voting rights are distributed equally amongst the members. The first two Baywind projects enabled a community in Cumbria to invest in local wind turbines. Baywind's aim is to promote the generation of renewable energy and energy conservation. Preference is shown for local investors, so that the community can share some of the economic benefits from their local wind farm. The wind represents an inexhaustible supply of 'free' energy and nobody actually owns the wind. Whoever owns the wind turbines receives the benefits from the sale of electricity that is produced. See www.baywind.co.uk. For more information on setting up community owned renewables projects see www.energy4all.co.uk.

Hayboxes: maintaining temperatures and storing food

It is very energy intensive to heat water and food, so once the energy has been used for heating it should be retained as long as possible. Using the same principle as a thermos flask, hayboxes are easy to make and a simple way to maintain temperature and significantly reduce cooking energy consumption. The box can be made of any material from cardboard to wood. Stronger and more durable materials will result in a better box. As with other simple technologies like solar cookers, the box is heavily insulated. Given the heavy insulation, when food is placed inside the box at temperature (for example, the pot of rice has reached boiling) the heat is retained and continues to cook the food. Remember that insulation maintains temperatures, hence it can also be used to keep its contents cold, so long as they are cold when placed inside.

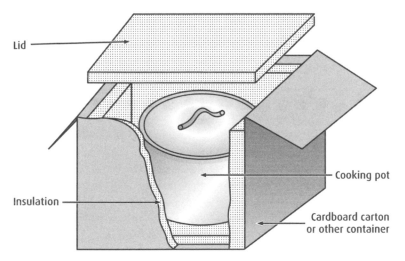

Lid

Cooking pot

Insulation

Cardboard carton
or other container

Figure 2.2   A haybox

Source: Adapted from Sunseed Desert Technology. Available at www.sunseed.org.uk

Solar showers

While 'civilisation' will not stand or fall based on our access to a nice warm shower, hot showers can be one of the most effective applications of appropriate solar technology. This technology is surprisingly effective and even in cloudy temperate areas there is an unexpected amount of solar energy, even in northern latitudes. The design uses a simple insulated tray that heats in the same way as a bucket of water when left in the sun. By adjusting the amount of insulation, the water depth in the tray, the use of reflectors and the proper sealing of the transparent cover – the heat capture can be maximised.

1. Make a rectangular frame out of wood (sheet metal, cement, cob or even just dig a hole in the ground).

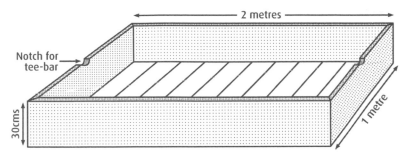

Figure 2.3   Solar shower 1

Source: Adapted from Peace Corps Guatemala, *A Guide to Appropriate Technology.*

Dimensions: 2 m long; 1 m wide; 30 cm high. (These dimensions will hold 172 litres. A more shallow design will heat up quicker but will not retain the heat as long.)

If the water heater is to be used on the ground then no bottom will be necessary; otherwise use plywood or planks for the bottom.

2. Optional but recommended – a reflector. This will help the system capture solar heat. A piece of sheet metal or plywood painted white and mounted behind the water heater will reflect more heat into the water.

3. Fit the tray with insulation to a depth of 15 cm – dry leaves, wood shavings, sawdust, fibreglass, polystyrene, etc.

Figure 2.4   Solar shower 2

Source: Adapted from Peace Corps Guatemala, *A Guide to Appropriate Technology.*

4. The water container should ideally be black sheet metal. Black to absorb the full spectrum of solar radiation and metal because in direct sunlight the temperatures will be hot – plastic could melt. If metal is not available and plastic or a similar material is used, then the areas that are not covered by water need to have a flap of white plastic or tinfoil over them to reflect as much solar radiation away from the material to avoid meltdown.

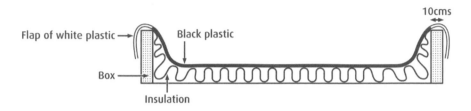

Figure 2.5   Solar shower 3

Source: Adapted from Peace Corps Guatemala, *A Guide to Appropriate Technology.*

5. Fill the box to a depth of 10–12 cm of water. Then lay clear plastic over the top allowing it to float on the surface of the water to prevent evaporation. It is important that the clear plastic be loose enough to rise and fall with a changing water level.

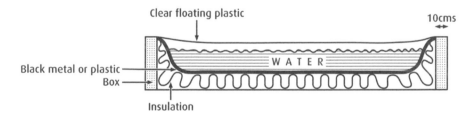

Figure 2.6   Solar shower 4

Source: Adapted from Peace Corps Guatemala, *A Guide to Appropriate Technology.*

6. The final step is to fit a transparent lid onto the box. If the lid is made of glass, then glue pieces of non-absorbent materials along the top edges of the frame to cushion the glass. A glass lid will be surprisingly heavy whereas a plastic lid will be quite light.

For glass:

☆ Place two aligned notches along the top edge and in the middle of the top and bottom pieces of the tee-bar to support the glass panes.

☆ Make notches in the top edges of the wooden frame to support crossbars of 5 cm metal across the box to support the glass panes.

☆ Place the eight glass panes onto the supporting tee- and crossbar framework.

☆ Seal the glass panes in their frame with white plastic tape.

If glass is unavailable or a simpler system is preferred, then clear plastic or vinyl can be stretched tightly over a crosswork of string. Leave at least a 2 cm air gap for insulation between the floating plastic and the transparent lid. This air space serves as insulation.

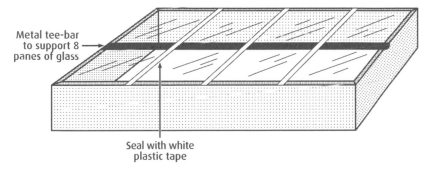

Metal tee-bar to support 8 panes of glass

Seal with white plastic tape

Figure 2.7   Solar shower 5

Source: Adapted from Peace Corps Guatemala, *A Guide to Appropriate Technology.*

7. The whole box is then tilted slightly towards the sun for the best solar exposure and so that rain will run off and heated water can be collected from the lowest point.

8. To collect the hot water you can:
    (a) Scoop the hot water out with a bucket; or
    (b) Place the heater on top of a frame or roof and remove the water through a tube. You can either fit a metal tube (with gaskets) at the low end of the tank – you will need some way (like a valve) of shutting off the water flow when desired – OR if you want to use a plastic tube, make a notch in the top of the

wood for the tube to sit in. Siphoning water out at the low end of the tank may be easier if plastic is used.

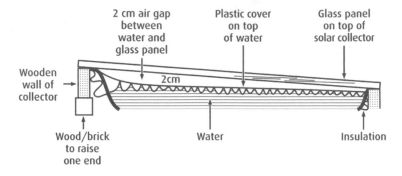

Figure 2.8   Solar shower 6

Source: Adapted from Peace Corps Guatemala, *A Guide to Appropriate Technology.*

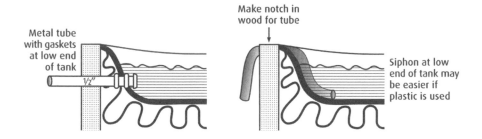

Figure 2.9   Solar shower 7

Source: Adapted from Peace Corps Guatemala, *A Guide to Appropriate Technology.*

The metal or plastic tube is also used for refilling the tray.

9. A complete system.

    ☆ The solar heater may also be placed on a roof or stand above the shower stall; this will allow gravity to feed the water down to the shower.

    ☆ The tube that carries the hot water from the collector to the shower can have either a shower head or simply holes put into the tube end to create the 'shower' effect.

★ For safety and maintenance reasons you should ideally have
   a valve (a component usually with a small lever or a screw
   that needs turning) that will shut off the water flow through
   the pipe.

★ A valve on the hot water from the collector to the shower is
   advisable and, if you have a cold water feed, a garden hose from the mains
   connecting into the shower line.

★ The cold water feed would also be used to refill the collector after hot water has
   been drawn off.

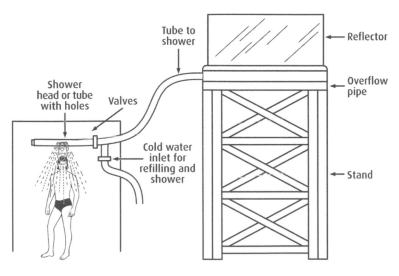

Figure 2.10   Solar shower 8

Source: Adapted from Peace Corps Guatemala, A Guide to Appropriate Technology.

waste

The by-product of lives that are always on the go is our huge consumption of *products*.
Many of the items we use on a daily basis are designed to be thrown away and
the energy and resources that went into their making with them. We are rapidly
approaching the point where landfills are full and the only option in some places is
waste incineration which is energy and emissions intensive. It was recently stated

that every second, householders in England bin almost four tonnes of waste, and we are throwing out 3 per cent more each year (*Guardian*, July 2006).

Box 2.3 Today...

Reduce, reuse, recycle, compost (see Chapters 9 and 10): refuse products with packaging; only buy products with less or no packaging; buy second-hand, donate clothes and household goods to friends, neighbours or second-hand shops; join the local 'Free-cycle' network; use old clothes as drying up clothes; reuse paper in the printer; refill your printer cartridge; refuse junk mail; recycle and salvage building materials; use reusable nappies; get a reusable menstrual product such as Mooncup; take a thermos and a packed lunch; take your cup/bowl to the takeaway; share a car/printer with friends or neighbours; get a wormery; use natural dyes and paints; tell your school, workplace, local cafe, shops, etc. to get their waste in order.

Compost toilets

Composting of human faeces is as ancient as digging a hole to shit in the woods. Many cultures continue composting today because the energy embodied in the faeces is an important resource in the sustainable production of agriculture. Into the modern day in China, many cities remain ringed by an area of peri-urban agriculture where the 'night soil' (faeces) is brought out of the city each morning to begin its reintroduction into the food cycle (although this is sadly changing quickly in the face of China's massive development boom of the past decade).

Composting toilets are an essential strategy for sustainability. The modern Western home uses upwards of 100,000 litres of fresh, drinkable water each year, just to flush the toilet. Composting toilets provide a solution that avoids using any water for flushing, does not produce sewage or contaminate ground water and saves nutrients in the faeces instead of 'putting them out to sea'. While the design shown in Figure 2.11 is a do-it-yourself model, there are much more modern and urban oriented designs available which include porcelain bowls and which can be designed within larger buildings in the city.

A composting toilet can be very hygienic and if properly operated does not smell. The most common model of compost toilet is designed with twin vaults. The vaults, usually made of brick or concrete, are used alternately. When one chamber is almost full, it is sealed off (a board bolted over the hole) and the other chamber is opened up. The first vault is left to compost while the other is in use. In temperate climates, usually a year is a sufficient length of time for good composting. The vaults are built with an access panel at the back so that the compost can be removed. The compost truly comes out looking like rich soil, with no odour or any other sign of the original material. Once the compost is removed the chamber is ready for another cycle and the vaults can be switched when appropriate.

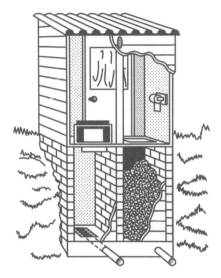

Figure 2.11   A two-chamber compost toilet

Source: Adapted from Intermediate Technology Group.
Available at
http://practicalaction.org/?id=technical_briefs_water

The key to a hygienic and smell free compost toilet is the addition of 'soak' material ensuring good ventilation. The microorganisms that do the composting work require the right balance of carbon and nitrogen in the environment to provide the elements they need to flourish. Since faeces and particularly urine (which inevitably ends up in the vault) is very nitrogen rich, it is essential to add carbon to the toilet after each use. There are a wide variety of soak materials – sawdust, leaves, dry cooking ashes, newspaper and straw can be used so long as they are rich in carbon – usually a handful of the material is sufficient. The soak material also helps to cover the faeces so that the flies are kept down and literally soaks up the excess liquid (from urine). While the finished compost should be human pathogen free, it is advisable never to use it directly on food crop soil but instead to use the finished compost on fruit trees, berry bushes and the like. Urine can also be siphoned separately and used as a compost activator.

There are a number of considerations when designing your compost toilet.

Ventilation and flies   Composting requires oxygen and therefore some kind of ventilation is essential. This issue has to be considered in context with the fly problem because any place where air can get in, so can flies. A good compost toilet will seal the vault with wire meshes for the ventilation pipe and seal under the toilet seat so that air can get in but flies cannot. Ideally, the toilet seat should have a tight seal (using glued foam and the like) so that no light or air enters the chamber and there should be a tightly screened ventilation hole in the back wall.

Urine   This can be a big problem if it puts too much liquid with high salt and nitrogen content into the vault. It is best to urinate somewhere else. However, it is inevitable that urine will find its way into the vault and therefore in addition to the soak material a drainage pipe (see Figure 2.11) and a slightly inclined vault floor, encouraging the liquid to run out of the chamber, are ideal.

Another urine solution is the straw bale urinal. This urinal is a very simple construction and makes excellent compost for your garden in three to four months depending on use. It consists of a bale of straw for men and women to urinate on. Men wee standing; a seat is constructed for women made from two thick pieces of wood (or a couple of bricks) across the bale where a toilet seat can rest on. The liquid soaks in and composts the middle of the bale. Try to place the bale in a place protected by rain.

If you want to have a compost toilet, but also want to be able to wee in it, there are solutions. You can separate the urine from the solids at the source, which means you have a smaller size vault and less hassle with draining. You can make a simple one yourself from a curved sheet of aluminium attached to the toilet pedestal at the point where you pee sitting. There are also designs that enable porcelain bowl lovers to have a flush toilet inside, but at the same time compost their waste. They separate urine from solids behind your house, where you also have your composting chamber. Information on these designs are in the 'water and waste' resources section at the end of this chapter.

water

Global water use has doubled since 1960 and as climate change looks set to have devastating impacts on our water cycles, creating hotter and drier climates, it becomes even more important to preserve this most precious of resources. Scarcity of clean water will also inevitably lead to more conflict over access to this resource. All these themes are interlinked – water is used in many industrial processes from making micro chips to printing paper and decontaminating paint from sewage, so reducing energy and waste will also impact on water use.

Box 2.4 Today...

Instead of wasting 4–5 litres of clean drinking water with every toilet flush, which equals about 150 drink cans per day, take a plastic bottle filled with water, put it in your toilet cistern to displace the water volume it occupies (1.5 or 2 litres) – now every time you flush you use only 2–3 litres; next time you have a shower, save the water, use this to flush the toilet; water the garden with old bath and shower water; store rainwater; brush teeth and shave with the tap turned off; take a shower not a bath (and a quick one at that); fix dripping taps; half fill the sink and use this water to rinse dishes or wash fruit and vegetables rather than leaving the tap running; install a rainwater collection barrel, use this for watering the garden and even flushing the toilet; use natural cleaning products and eco paints; avoid pouring out of date milk, juice or food down the drain, saving energy in the sewage treatment process and reducing the BOD (biological oxygen demand) for the breakdown of these particles.

Grey water system

Any water that has been used in the home, except water from toilets, is called grey water. Dish, shower, sink and laundry water comprise 50–80 per cent of residential 'waste' water. This may be reused for other purposes, especially landscape irrigation.

Although there are a number of commercially available grey water systems it is also possible to make one. The basic idea is to pass the grey water through a series of tanks in order to filter and clean the water so that it can be reused in the garden or go back into a river. It is important to use ecological and biodegradable soaps and washing-up liquids in a grey water system. See the resource section for more information.

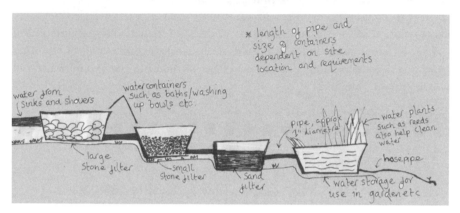

Figure 2.12   Grey water system

Source: Kim Bryan.

Water filtering at home

While getting fresh water is very important (clean drinking water and basic sewage services prevent many diseases), having the technology to clean water is even more so. This technology can be very basic and relatively cheap to implement on a family or community scale. It can be useful in emergency or temporary situations. But be warned: the consequences of a poorly constructed or maintained filtering system can be very sick people – do your research first!

Filtering water through sand removes most and sometimes virtually all of the solid and organic impurities present in the water. The sand filtration systems described here are suitable for eliminating solids and biological organisms but they cannot cleanse water with chemical impurities or condition water for high or low pH levels.

The water passes from the top of a tank to the bottom with suspended particles being sieved out by the sand. A bio-film of microorganisms will naturally develop on top of the first few centimetres of sand grains that do much of the work to purify the water. A sand filtration system will not work properly without this bio-film. Wait a minimum of two days before using the water from the system.

The water level in the tank can be controlled automatically by
a ballcock that turns off the water flow; otherwise you need to fill
the tank yourself. Usually there is 20–40 cm of standing water,
followed by 50–150 cm of sand, provided that the sand grain size
is between 0.15 and 0.3 mm in diameter. After the sand, 10–30 cm
of pea gravel is set at the bottom of the container. If possible a permeable geo-textile
layer such as a nylon curtain should separate the sand and gravel layers to prevent
sand clogging the outlet pipe or other system components down the line. A slotted
pipe or pipe with holes drilled in it takes the filtered water from the tank out to your
water supply. Both the sand and gravel should be washed before installation into a
sand filter to remove any remaining silt.

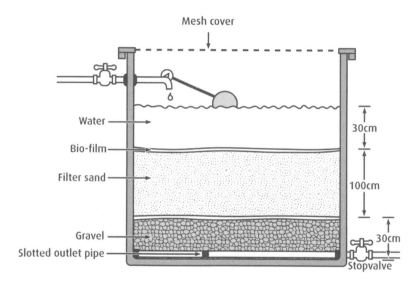

Figure 2.13   Slow sand filtration at home

Source: Adapted from Thornton (2005). 'Slow Sand Filters Tipsheet'. *The Water Book.*

To provide good treatment the water flow should be no more than 100 litres of
water per square metre of filtering sand per hour (100 L/m²/hr). The rate of water
outflow can be controlled by the stop valve on the outlet pipe. To design a home
system you need to record the litres of filtered quality water that you need each day
and size your system accordingly.

Sand filters can be made of any kind of tank such as polyethylene, concrete or
rendered block work. However the internal sides of the tank need to be very rough,

otherwise water will slide down the smooth sides and avoid filtration in the sand. The tank will need a cover to prevent bird droppings and leaves getting in. In temperate climates allowing UV penetration into the tank helps with the sterilisation process. In hot climates no light of any kind can be allowed, otherwise this encourages algae growth which is not desirable.

Like any other system, a slow sand filtration system requires maintenance: the system will need draining for cleaning every three to four months where the top 5 cm of sand is removed entirely or rinsed before being returned to the filter. If there is a frost risk, a tap must be kept running down-line at all times so that water does not freeze inside the pipe.

prospects for the future

While the consuming classes in the 'developed' world are entertained by the latest MP3 players and pet-replicating robots, there are 1.3 billion human beings – roughly a quarter of all humanity – who do not have access to basic electrical services for lighting in their homes. There are vast numbers of people who live in absolute poverty where food and water is a daily struggle.

The cost of bringing people out of poverty and providing basic health services for every living person is minuscule compared with the cost of manufacturing more military arms and subsidising oil production. There is a massive movement growing across the world of self-help and solidarity organisations that are engaging in 'development from below'. Appropriate technology is an important element in this recipe for community autonomy and self-governance, as well as an important form of self-sufficiency and insurance if, or when, the present global order collapses. This chapter has shown some ways that we can begin to be part of this movement, with simple things we can do today and more ambitious and creative schemes we can embark upon with friends and family. These simple, painless adjustments in our lives can inspire us to think about some of the larger, more challenging ones.

Bryce Gilroy-Scott has worked and studied at the Centre for Alternative Technology (CAT), Wales. He is currently involved in a number of renewable energy and sustainability community projects and building an ecovillage in the East Midlands of England in a woodland with a social enterprise that works with disengaged youth. Additional material on grey water filtration was provided by Starhawk, writer, activist, trainer and permaculture designer, and Malamo Korbetis, student and long-term volunteer at CAT.

resources

Books, guides and reports

Industrial society, human civilisation and ecological crisis

Bookchin, Murray (2004). *Post-scarcity Anarchism*. Edinburgh: AK Press.

Bronowski, J. (1972). *The Ascent of Man*. BBC series.

Carson, Rachel (1963). *Silent Spring*. London: H. Hamilton.

Diamond, Jared (1997). *Guns, Germs and Steel: A Short History of Everybody for the Last 13,000 Years*. New York: Jonathan Cape.

Diamond, Jared (2005). *Collapse: How Societies Choose to Fail or Survive*. New York: John Allen.

Harris, Marvin (1977). *Cannibals and Kings: The Origins of Cultures*. New York: Random House.

Heinberg, Richard (2005). *The Party's Over*. Forest Row, Sussex: Clairview Publishers.

Hill, Christopher (1972). *The World Turned Upside Down: Radical Ideas during the English Revolution*. London: Maurice Temple Smith.

Jensen, Derek (2004). *A Language Older than Words*. New York: Context.

Kovel, Joel (2002). *The Enemy of Nature. The End of Capitalism or the End of the World*. London and New York: Zed Books.

Meadows, Donella et al. (1972). *The Limits to Growth: A Report for the Club of Rome's Project on the Predicament of Mankind*. London: Earth Island Publishing.

Monbiot, George (2000). *Captive State: The Corporate Takeover of Britain*. Basingstoke: Macmillan.

Mumford, Lewis (1966). *The Myth of the Machine*. New York: Harcourt Brace Jovanovich.

Norberg-Lodge, Helena (1992). *Ancient Futures: Learning from Ladakh*. London: Rider.

Pilger, John (2002). *The New Rulers of the World*. London: Verso.

Sahlins, M. (1974). *Stone Age Economics*. London: Tavistock.

Sale, Kirkpatrick (1995). *Rebels Against the Future. The Luddites and Their War on the Industrial Revolution: Lessons for the Computer Age*. Reading, Mass.: Addison Wesley.

Schumacher, E.F. (1973). *Small is Beautiful: A Study of Economics as if People Mattered*. London: Blond and Briggs.

Stern, Nicholas (2007). *The Economics of Climate Change: The Stern Review*. Cambridge: Cambridge University Press.

Tainter, Joseph (1988). *The Collapse of Complex Societies*. Cambridge and New York: Cambridge University Press.

Thompson, E.P. (1968). *The Making of the English Working Class*. Harmondsworth: Penguin.

World Commission on Environment and Development (1987). *Our Common Future.* Oxford: Oxford University Press.

Zerzan, J. (1996). *Future Primitive and Other Essays.* Brooklyn, NY: Semiotext Books.

Sustainable communities

Heinberg, Richard (2004). *Power Down: Options and Actions for a Post Carbon World.* Gabriola Island, BC: Clairview Books.

Nozick, Marcia (1992). *No Place Like Home: Building Sustainable Communities.* Ottawa: Canadian Council on Social Development.

Seymour, John (1976). *The Complete Book of Self-Sufficiency.* London: Corgi Publishing.

Weisman, Alan (1998). *Gaviotas: A Village to Reinvent the World.* White River Junction, VT: Chelsea Green Publishers.

Sustainable economics

Douthwaite, Richard (1999). *The Growth Illusion: How Economic Growth has Enriched the Few, Impoverished the Many and Endangered the Planet.* Gabriola Island, BC: New Society.

Schumacher, E.F. (1973). *Small is Beautiful: A Study of Economics as if People Mattered.* London: Blond and Briggs.

Sirolli, Ernest (1999). *Ripples from the Zambezi. Passion, Entrepreneurship, and the Rebirth of Local Economies.* Gabriola Island, BC: New Society Publishing.

Sustainable construction

Borer, Pat and Cindy Harris (2005). *The Whole House Book.* Machynlleth: Centre for Alternative Technology Publications.

Kahn, Loyg (2004). *Home Work: Handbuilt.* Bolinas, CA: Shelter Publications.

Stulz, Roland and Kiran Mukerji (1998). *Appropriate Building Materials: A Catalogue of Potential Solutions.* London: ITDG Publishing.

Sustainable sewage and clean water

Del Porto, David and Carol Steinfeld (2000). *The Composting Toilet System Book: A Practical Guide.* Center for Ecological Pollution Prevention. White River Junction, VT: Chelsea Green Publishers.

Grant, Nick, Mark Moodie and Chris Weedon (2005). *Sewage Solutions: Answering the Call of Nature.* Machynlleth: Centre for Alternative Technology Publications.

Harper, Peter and Louise Halestrap (2001). *Lifting the Lid: An Ecological Approach to Toilet Systems.* Machynlleth: Centre for Alternative Technology Publications.

Slow Sand Filters. Tipsheet. Machynlleth: Centre for Alternative Technology Publications.

Thornton, Judith (2005). *The Water Book: Find it, Move it, Store it, Clean it ... Use it*. Machynlleth: Centre for Alternative Technology Publications.

Wind micro generation

Piggott, Hugh (1997). *Windpower workshop. Building Your Own Wind Turbine*. Machynlleth: Centre for Alternative Technology Publications.

Piggott, Hugh (2004). *It's a Breeze. A Guide to Choosing Windpower*. Machynlleth: Centre for Alternative Technology Publications.

Permaculture and sustainability

Burnett, Graham (2001). *Permaculture Beginners Guide, Spiralseeds*. East Meon, Hampshire: Permanent Publications. (A great little book, beautifully illustrated from the world's greatest punk permaculturalist.)

Gilman, Robert and Diane (1991). *Ecovillages and Sustainable Communities*. Bainbridge: Gaia Trust.

Hemenway, Toby (2001). *Gaia's Garden. A Guide to Home-Scale Permaculture*. White River Junction, VT: Chelsea Green Publishing.

Holmgren, David (2002). *Permaculture: Principles and Pathways Beyond Sustainability*. Melbourne, Australia: Holmgren Design Services. (Philosophy and principles of permaculture.)

Jenkins, Joseph (1999). *The Humanure Handbook*. White River Junction, VT: Chelsea Green Publishing.

Lyle, J. (1997). *Regenerative Design for Sustainable Development*. London: John Wiley.

Mollison, Bill (1997). *Permaculture: A Designer's Manual*. Tyalgum, Australia, Tagari. (The classic, comprehensive book on the subject.)

Starhawk (2002). *Webs of Power*. Gabriola Island, BC: New Society Publishers. (Essays and reports on the global justice movement.)

Starhawk (2004). *The Earth Path*. San Francisco: HarperSanFrancisco. (Permaculture and earth based spirituality, a good introduction to both practical and mystical practice.)

Whitefield, Patrick (2004). *The Earthcare Manual*. Hampshire: Permanent Publications. (Great permaculture resource for the UK and cold temperate climates.)

Woodrow, Linda (1996). *The Permaculture Home Garden*. Ringwood, Victoria, Australia: Viking Penguin. (A great resource for the beginner and for home-scale permaculture.)

Websites

General

The Camp for Climate Action www.climatecamp.org.uk/links.htm

Climate Change news and action www.climateimc.org and www.risingtide.org.uk

Climate Outreach and Information Network www.coinet.org.uk
Earth Policy Institute www.earth-policy.org/
Facts sheets on energy, water, waste, transport www.lowimpact.org
Otesha Project Resource Guide www.otesha.ca/files/the_otesha_book. pdf
Peak oil www.peakoil.net

Permaculture

Earth activist training www.earthactivisttraining.org
Permaculture Association (Britain) www.permaculture.org.uk.
Permaculture Magazine www.permaculture.co.uk
Regenerative Design Institute www.regenerativedesign.org
Starhawk's writing www.starhawk.org

Eco-living and appropriate technologies

Alternative Technology Association http://ata.org.au
Autonomous Coordinating Group of Appropriate Technology for Health www.catas1.
   org
Builders without Borders advocating natural building techniques http://builderswith-
   outborders.org/
Centre for Alternative Technology www.cat.org.uk
Feasta: Foundation for the Economics of Sustainability www.feasta.org
Global Ecovillage Network http://gen.ecovillage.org/
Practical Action (formerly the Intermediate Technology Development Group) www.prac-
   ticalaction.org
Yellow house guide to eco-renovation www.theyellowhouse.org.uk

Energy

Baywind Energy Co-operative www.baywind.co.uk
Centre for Sustainable Energy www.cse.org.uk
Energy descent www.transitionculture.org
Energy4All www.energy4all.co.uk
Guides on how to harness the power and heat of the sun www.builditsolar.com
Learning to live in a low energy world www.postcarbon.org
Renewable energy courses in Spain www.escanda.org
Renewable energy courses in the UK www.greendragonenergy.co.uk
Sunseed Desert Technology www.sunseed.org.uk
World Alliance for Decentralized Energy (WADE) www.localpower.org

Water and waste

Action on the 3 Rs (reduce, reuse, recycle) www.wastewatch.org.uk
Elemental Solutions Compost Toilets www.elementalsolutions.co.uk/

Everything you need to know about compost toilets www.
   compostingtoilet.org

Grey water resources and courses www.oasisdesign.net

Intermediate Technology Development Group: Compost Toilets.
   Technical brief http://practicalaction.org/docs/technical_
   information_service/compost_toilets.pdf

Natsol Compost Toilets www.natsol.co.uk/

Transport

Bike tours, education and entertainment www.bicycology.org.uk

Campaign for sustainably sourced biofuels www.biofuelwatch.org.uk

Campaigning against road building www.roadblock.org.uk

Direct Action opposing airport expansion and short-haul flights www.planestupid.com

How and why to travel without flying www.noflying.info

Sustainable transport and cycling promotion www.sustrans.org.uk

3 why do it without leaders

The Seeds for Change Collective

The alarm clock rings. Shower, dress, listen to the news. Get irate: war in Iraq – no one's asked me! Tax increase, great. Yet another step closer to privatising the health service. Local elections coming up, politicians make new wonderful promises. Why bother? Rush to work, another dull day in the office. Get called in by the boss – new targets from head office, work overtime this week. That's my day off gone. Get home, microwave some food. Letter from the landlord: pay more or move out. Too tired to go out, just switch the television on for some light relief. Had a good day?

This chapter is about how we relate to each other and how we organise society. We are all, to some extent, controlled by others who don't understand or care about our wants and needs – managers, landlords, city councils, creditors, police, courts, politicians. And all of us exert power over others in varying degrees – in the home, at work, at school. How do we break out of this system of control, where we all, willingly or unwillingly, exert power over others, forcing them into actions they'd rather not do?

One solution is to challenge and provide alternatives to the rules, leaders and hierarchies that largely direct our daily lives and shape the way our societies function. We need to develop a different understanding of power – where people work with each other rather than seeking to control and command. And we need to find ways of relating to each other without hierarchy and leaders. These ideas are far from new and this chapter is part of a journey into a different world, where people have always striven for control over their own lives, struggled for self-determination and to rid themselves of their rulers and leaders. At the core of these struggles for liberty lies the desire of every human being to live a fulfilled life, following her interests, fulfilling her needs. A desire that extends towards creating a society where this is possible not just for a few, but for everyone. What follows is an exploration of ways of making decisions collectively and why it's important to organise society without leaders.

what's wrong with leaders?

> We all know that happiness comes from control over our own lives, not other people's lives. (CrimethInc. 2000, 42)

Many of us have been brought up in a culture which believes that Western-style democracy with one-person-one-vote and elected leaders is the highest form of democracy. Yet in the very nations which shout loudest about the virtues of democracy, many people don't even bother voting anymore. They feel it doesn't make any difference to their lives.

When people vote for an executive they also hand over their power to representatives to make decisions and to effect change. Representative democracies create a system of hierarchy, where most of the power lies with a small group of decision makers on top and a broad base of people whose decisions are made for them at the bottom. People are often inactive in this system because they feel that they have no power and that their voice won't be listened to. Being allowed to vote 15 times in our lives for an MP or senator is a poor substitute for making decisions ourselves.

Even though our government may call itself democratic, there are many areas of our society where democratic principles have little influence. Most institutions and workplaces are hierarchical: students and employees don't usually get a chance to vote their superiors into office or have any decision-making power in the places where they spend the greatest part of their lives. Or consider the supermarket chain muscling its way into a town against the will of local people. Most areas of society are ruled by power, status and money, not democracy.

A desire for something different is nothing new. People have been refusing to accept the 'god given' world order and struggled for control over their own destiny in every society humanity has known.

taking back control

> We have these moments of non-capitalist, non-coercive, non-hierarchical interaction in our lives constantly, and these are the times when we most enjoy the company of others, when we get the most out of other people; but somehow it doesn't occur to us to demand that our society works this way. (CrimethInc. 2005)

The alternatives to the current system are already here, growing in the gaps between the pavement stones of state authority and corporate control. We only need to learn to recognise them as seedlings of a different kind of society. Homeless people occupying empty houses and turning them into collective homes, workers buying out the businesses they work for and running them on equitable terms, friends organising a camping trip, allotment groups growing vegetables on patches of land collectively; once we start looking there are hundreds of examples of co-operative organising that we encounter in our daily lives. Most of these organise through varying forms of direct democracy. Direct democracy is the idea that people should have control over their lives, that power should be shared by all rather than concentrated in the hands of a few. It implies wide-ranging liberty, including the freedom to decide one's own course in life and the right to play an equal role in forging a common destiny.

This ideal is based on two notions: first, that every person has the right to self-determination, the right to control their own destiny and no one should have the power to force them into something; and second, that as human beings most of us wish to live in society, to interact with other people. Direct-democratic systems aim to find a way of balancing individual needs and desires with the need for co-operation. Two forms of these systems are direct voting and consensus decision making.

Direct voting

It is only because people are not claiming their own power, because they are giving it away, that others can claim it for their own.

Direct voting does away with the need for leaders and structures of control. Decisions are made through a direct vote by the people affected by them. This ensures that decision-making power is distributed equally without giving group members absolute vetoes. When group members disagree, majority rule provides a way to come to a decision.

One of the problems with this is that the will of the majority is seen as the will of the whole group, with the minority expected to accept and carry out the decision, even if it is against their own needs, beliefs and desires. Another problem is that of a group splintering into blocs of different interests. In such cases decision making can become highly competitive, where one group's victory is the other group's defeat.

On the odd occasion people may find that acceptable, but when people find themselves in a minority they lose control over their own lives. It undermines commitment to the group and to the decisions taken. This often leads to passive membership or even splits in the group. Many groups using direct voting are aware of this problem and attempt to balance voting with respect for people's needs and desires,

Book
Depository

Book Depository

This #BDDesign bookmark is part of a set designed by our customers.

3/10

Bookmark designed by
Samantha Cheung
Hong Kong

Samantha's favourite book
The Little Prince
Antoine de Saint-Exupéry

View the full set:
bookdepository.com/bookmarks

spending more time on finding solutions that everyone can vote for, or proactively protecting minority interests.

Consensus decision making

No one is more qualified than you are to decide what your life will be.
Another form of direct democracy is making decisions by consensus. At its core is a commitment to find solutions that are acceptable to all. Instead of voting for an item consensus works creatively to take into account everyone's needs. Consensus is about finding common ground with decisions reached in a *dialogue between equals*, who take each other seriously and who recognise each other's equal rights. No decision will be made against the express will of an individual or a minority. Instead the group constantly adapts to all its members' needs.

In consensus, every person has the power to make changes in the system, and to prevent changes that they find unacceptable. The right to *veto* a decision means that minorities cannot just be ignored, but creative solutions will have to be found to deal with their concerns.

Consensus is about participation and equalising power. It can also be a very powerful process for building communities and empowering individuals. Another benefit of consensus is that all members can agree to the final decision and therefore are much more committed to actually turning this decision into reality.

Consensus can work in all types of settings: small voluntary groups, local communities, businesses, even whole nations and territories:

☆ Non-hierarchical societies have existed in North America for hundreds of years. One example is the Muscogee (Creek) Nation, where in those situations when consensus could not be achieved, people were free to move and set up their own community with the support – not the enmity – of the town they were leaving.

☆ Many housing co-operatives and social enterprises use consensus successfully: a prominent example is Radical Routes, a network of housing co-operatives and workers' co-operatives in the UK, who all use consensus decision making.

☆ The business meetings of the Religious Society of Friends (Quakers) use consensus to integrate the insights of each individual, arriving at the best possible approximation of the truth.

☆ Many activists working for peace, the environment and social justice regard consensus as essential to their work. They believe that the methods for achieving change need to match their goals and visions of a free, non-violent, egalitarian society. In protests around the world many mass actions involving several thousand people have been planned and carried out using consensus.

Different processes have developed both for small and larger groups of people, such as splitting into smaller units for discussion and decision making with constant exchange and feedback between the different units. However, like any method of decision making, consensus has many problems which need to be looked at.

★ As in any discussion those with more experience of the process can manipulate the outcome.

★ There can be a bias towards the status quo: even if most members are ready for a change, existing policies remain in place if no decision is reached.

★ Sometimes it can take a long time to look at ideas until all objections are resolved – leading to frustration and weaker commitment to the group.

★ The right to veto can be a lethal tool in the hands of those used to more than their fair share of power and attention. It can magnify their voices, and be used to guard against changes that might affect their power base and influence.

★ Those who do more work or know more about an issue will have more power in a group whether they like it or not. This is a two-way process – people can only dominate a group if others let them.

★ Where people are not united by a common aim they will find it difficult to come to the deep understanding and respect necessary for consensus.

Most of these problems stem from lack of experience in consensus rather than being inherent to the process. It takes time to unlearn the patterns of behaviour we have been brought up to accept as the norm. Living without hierarchy does get easier with practice!

Box 3.1   Consensus ≠ veto power

Unlike 'veto power' decision rule, consensus is based on the desire to find common ground. The veto power model, used in the UN Security Council and in parts of the European Union, works on mutual distrust and an unwillingness to compromise. The motivation behind negotiations is to prevent deadlock rather than to create a sense of shared goals and mutual respect.

creating societies without leaders

A society which organises itself without authority is always in existence, like a seed beneath the snow, waiting for a breath of spring air to rise up in its full beauty.
Alternatives to the current system of decision making in our society exist. We need to extend these spheres of free action and mutual aid until they make up most of society. It is the myriad of small groups organising for social change that will, when connected to each other, transform society. Once we realise that it is within our power to shape our environment and societies we can claim a new destiny for ourselves, both individually and collectively. In this we are only limited by our courage to imagine what can be, and by our willingness to learn how to coexist and collaborate. Societies based on the principle of mutual aid and self-organisation are possible. They have existed in the past and exist today. Our challenge is to develop systems for decision making that remain true to the spirit of self-government and at the same time allow decisions to be made that not only affect 20, 50, 200 people, but potentially tens or hundreds of thousands of people.

Self-government

> Every kind of human activity should begin from what is local and immediate, should link in a network with no centre and no directing agency, hiving off new cells as the originals grow. (Ward 1988, 10)

Self-government is based on the ideal that every person should have control over their own destiny. This ideal requires us to find ways to organise a society in which we can coexist with each other whilst respecting people's individuality, their diverse needs and desires. Direct democracy in small groups depends on group members sharing a common goal, building trust and respect, active participation, a clear process. Clearly these same conditions also need to apply to making decisions on a much larger scale. But when it comes to organising large groups (such as neighbourhoods, cities, regions or even continents) the following points are particularly important:

(a) Decentralisation
Decisions should be made by those that are affected by them. Only those with a legitimate interest in a decision should have an input. The more local, the more decentralised we can make decisions, and the more control we will each gain over our lives.

(b) Diversity is our strength

We all have different needs and desires. To accommodate these we need to create a fluid society full of diversity, allowing each to find their niche – creating a richly patterned quilt rather than forcing people into the same bland uniform. The more complex the society we create, the more stable it will be.

(c) Clear and understandable structures

While we need the fabric of our society to be complex, we want the structures of organising and making decisions to be simple and understandable. It needs to be easy for people to engage in decision making.

(d) Accountability

Being accountable means taking responsibility for your actions. This makes it more difficult to accumulate power and avoids corruption – common pitfalls of organising on any scale.

In practice this means developing a decentralised society, with decisions being made at the local level by the groups of people affected by them. These groups will be constantly changing and adapting to serve the needs of the people connected to them. Where we need co-operation on a larger scale groups can make voluntary agreements within networks and federalist associations. If the processes are easily understood, transparent and open, then accountability is added to the whole process.

So what would this society look like? How will services be organised, limited goods distributed, conflicts resolved? How can health care, public transport, the postal service be organised?

Neighbourhoods and workers' collectives – a federalist model

One model for structuring society is using neighbourhoods and workers' collectives as the two basic units for decision making. Within the neighbourhoods people co-operate to provide themselves with services such as food distribution and waste disposal. Workers' collectives work together on projects such as running a bus service, factories, shops, hospitals. Decisions in all these groups are made by direct democracy, each member being directly involved in making the decisions affecting their lives. Some of these groups vote, others operate by consensus but all are characterised by respect for the individual and the desire to find solutions that are agreeable to all. It may sound as if we have to spend all our time in committees and meetings, but in reality most things are worked out through informal and spontaneous discussion and co-operation: organising on a local level is made much easier through daily personal contact.

A lot of co-operation is required between all these collectives and neighbour-hoods. Working groups and spokescouncils bring together delegates from different interest groups to negotiate and agree ways of co-operating on a local, regional and even continental level. Not everyone has to go to every meeting – an efficient and sensitive communications network is developed between all groups and communities. This involves sending recallable and directly responsible delegates to meetings with other groups. These delegates can either be empowered to make decisions on behalf of the group or they might have to go back to their group to check for agreement before any decision is made. Decision making is focused on the local level, with progressively less need to co-operate as the geographical area becomes larger. The details are resolved locally, only the larger, wider discussions need to be taken to regional or inter-regional levels.

Box 3.2   Participatory budgets

Participatory budgets are a process of democratic deliberation and decision making, in which ordinary city residents decide how to allocate part of a public budget. In 1989 the first participatory budgeting process started in the city of Porto Alegre, Brazil. In a series of neighbourhood, regional, and citywide assemblies, residents and elected budget delegates identified spending priorities and voted on which priorities to implement. Participatory budgeting is usually characterised by several basic design features: identification of spending priorities by community members, election of budget delegates to represent different communities, facilitation and technical assistance by public employees, local and higher level assemblies to deliberate and vote on spending priorities, and the implementation of local direct-impact community projects. Since their inception in Porto Alegre the concept of participatory budgeting has spread to many other municipalities across the world.

Source: Adapted from Wikipedia and Participatory Budgeting.

Can it be done?

You might find it hard to imagine how collective services such as train travel or bus services through several communities can be organised without a central authority, particularly if each community is independent and answerable to its residents rather than a central government. But consider present day international postal services, or cross border train travel, which are organised across countries without a central authority. These are based on voluntary agreements – it is in everyone's interest to co-operate.

Throughout history there are many examples of people organising society themselves. Often this happens in those rare moments when a popular uprising withdraws support (and thus authority) from the state. This leaves a vacuum of power – suddenly it becomes possible for ordinary people to put ideals of self-government and mutual aid into practice on a larger scale.

The economic crisis of December 2001 in Argentina brought about a popular uprising that is still going on today. The gap left behind as the government lapsed into chaos and the local currency collapsed was filled by local people getting to know and supporting each other. Factories were squatted and owners evicted so that the collective could benefit from their own labours. Land was seized to grow food for the community. But perhaps the most interesting development was in the way people began to experiment with different ways of organising themselves, their workplaces and their communities. Traditional hierarchies have been abandoned as people become more confident in their own skills and in their rejection of government and bosses.

The remarkable events of the Spanish Revolution in 1936 were the culmination of decades of popular education and agitation. During the civil war, large parts of the country were organised in decentralised and collective ways. A famous example is the Barcelona General Tramway Co. which was deserted by its managers. The 7000 workers took over the running of the trams, with different collectives running the trams for different parts of the city. Citywide services were maintained by federalist co-ordination. The increased efficiency of the collectives led to an operating surplus, despite running more trams, cutting fares, increasing wages and new equipment! The general spirit was one of optimism and freedom.

Building a community based on voluntary networks and mutual aid

What follows are two case studies of contemporary self-organisation and voluntary association.

Case study 1: HoriZone ecovillage. A temporary village in resistance to the G8 summit, July 2005, Scotland   A recent example of people creating a society based on co-operation is the ecovillage in Stirling, Scotland. Having come together with the aim of protesting against the Group of 8 nations summit and the global power system it represents, the people living in the ecovillage were also aiming to experiment with, and experience, a free society. For ten days, 5000 people from different parts of the world lived together communally in a tented, temporary village and put their ideas into practice. The ecovillage offered a unique chance to experiment with consensus decision making on a large scale. This was particularly exciting as one of the criticisms always levelled at consensus is that it might work for 20 people but that it would be impossible to organise whole communities or even countries on this basis.

At the heart of the village were neighbourhoods of 50–200 people, where people lived, ate, discussed and relaxed together. Most neighbourhoods were based on geographical areas that people had come from (such as Manchester neighbourhood), others were based on shared interests (such as the Queer neighbourhood). People either arrived as part of a neighbourhood or joined one to their liking. Life in the neighbourhood was organised collectively, with shared meeting spaces, communal food, water and toilets. Work was done voluntarily, with the ideal that it would be shared out equally amongst everyone.

Working groups from different neighbourhoods with relevant skills and interests were set up and co-ordinated these activities. This included buying and distributing food, maintaining the water and grey water systems, first aid/medical care, camp-wide health and safety, refuse collection, and transport to and from the camp. Delegates from all working groups and all neighbourhoods met daily in the format of a spokescouncil for a site meeting, where this work was co-ordinated, policies agreed, and jobs identified and allocated. Delegates were generally rotated from day to day, were accountable to their groups and had limited decision-making power. Generally this worked well, everybody had enough to eat, enough water to drink and wash with, and a place to sleep in.

> 'Most people find it hard to imagine a whole society based on free association and co-operation, since most of us have only experienced societies based on hierarchy and competition. This is what was so amazing about the Ecovillage in Stirling. It was possible to catch glimpses of what a free society could be like: so many moments of co-operation, of people helping each other to overcome adverse circumstances'.
> (Participant at Stirling HoriZone)

There were a number of key challenges. First, while on a daily basis thousands of people took part in meetings both on a neighbourhood and site level, it was really difficult involving not just the majority of people but everyone. Some had no idea of how the camp worked, while others were busy organising actions or maintaining essential infrastructures. A facilitation group was formed and worked hard to make processes transparent and to involve everyone in the decision-making process. A second challenge we faced was balancing our own desires with the needs of our neighbours, especially in terms of setting agreed rules for things like quiet times and music volumes.

Case study 2: Zapatista autonomous municipalities   Since their uprising in January 1994 in the Mexican state of Chiapas, the Zapatista movement has been quietly building a parallel system of government based on local autonomy – linking present politics to traditional ways of organising life in indigenous communities. The Zapatista system of 'good government' contrasts sharply with what they call the 'bad government' of official representational politics in Mexico City. Zapatista villages are clustered into autonomous municipalities. These are run by an autonomous council (*consejo autonoma*) and everyone has to take turns in running them. In turn, clusters of about six municipalities form Good Government Juntas in a particular region (which acts like a mini-parliament). These juntas are based in physical places called the 'Caracoles' (which act like mini town halls) and form the first point of contact for the outside world.

The main function of the juntas is to counteract unbalanced development and mediate conflicts between the autonomous communities. Each junta also levies a 'brother tax' of 10 per cent of the total costs of all external projects undertaken in their zone which helps pay for the expenses of the junta. The juntas also organise rotas of volunteer interns to run the zone hospitals, schools and workshops. What makes this system of government special is that it is based on rotation of the delegates – it is not the people or personalities that endure but the functions they fulfil and pass on to others. The delegates have to learn how to govern and pass on the collective knowledge and information to the next team, which means that more knowledge and skills are spread throughout the community. At the heart of the juntas is the Zapatista idea of 'governing by obeying' – that governing is about listening and responding, not dictating, and that if people govern poorly they are recalled immediately. It all sounds complex and at times it is. The fact that everyone takes part often makes it confusing and slow and means there is less consistency. But this is real democracy in action where everyone takes part.

These case studies highlight areas we need to continue to develop:

(a) The first issue is a wider one around balancing our own desires with the needs of others. If we are to be free to make our own choices this will sometimes impact on what others can and can't do. The concept of having a multitude of different neighbourhoods and working collectives from which to choose from helps in this context: what is socially acceptable will be different in each neighbourhood. People will choose their place to live with that in mind. However if you can't fit in with your neighbours, it is not always easy or practical to move away. And we don't want to create lots of mini-ghettos which don't communicate. We need to find effective ways of resolving such conflicts without recourse to a 'higher authority' even in a diverse society. The next chapter on consensus decision making outlines some practical ways of dealing with this problem.

(b) The second issue is about how we make decisions that involve many different groups. Not everyone can be in each meeting at the same time (nor would they want to be!). We need to find effective and simple ways to delegate and make decisions on a large scale. The spokescouncil is one option and is explained in more detail in the next chapter. But we need to work hard, as the Zapatistas have done, to ensure openness and accountability – especially when the spokescouncil consists of thousands of people and there are several tiers of delegates. Experience tells us we need to develop ways of delegating, learning to trust each other and also how to take account of the needs and views of those not present when making decisions. We may be able to combine concepts such as spokescouncils and making decisions online to provide an answer to the challenges posed by large-scale consensus-based decision making.

turning our dreams into reality

Let us put this ideal – no masters, no slaves – into effect in our daily lives however we can, creating glimpses of free society in the here and now instead of dreaming of a distant utopian age.

In this chapter we've looked at how society might be organised more equitably. But these ideas aren't going to become reality by magic. The case studies and examples show that people have been doing it without leaders in many places around the world. It's up to all of us to learn the lessons from these experiences and apply them to organising our daily lives, our neighbourhoods and places of work. We need to continue to come up with creative solutions to the challenges that working without leaders throws up. Above all we need to share and build on our experiences of doing it without leaders, helping us to avoid creating new forms of hierarchy and control.

This need for research and skill sharing on making decisions without leaders has given rise to training collectives such as RANT in the USA and Seeds for Change in the UK. Such collectives are themselves examples of self-help and mutual aid where, based on their own experience, members offer free workshops, resources and advice to community and action groups. Everyone has skills that are worthwhile sharing with others. Here are eight steps that you can take for gaining control over your life:

- ☆ Get to know your needs and desires and learn to express them.
- ☆ Learn to understand and respect the needs and desires of others.
- ☆ Refuse to exert power over others. Look at your relationships with your family, friends and colleagues.
- ☆ Start organising collectively and without hierarchy – in community groups, in unions, at work.
- ☆ Start to say no when your boss is making unreasonable demands. Stop making demands of others.
- ☆ Learn about power and the true meaning of democracy. Get to grips with the ins and outs of consensus decision making.
- ☆ Share your knowledge and skills with the people around you.
- ☆ Don't give up when the going gets rough. Work out what's going wrong, make changes, experiment.

4 how to make decisions by consensus

The Seeds for Change Collective

Chapter 3 looked at different ways of making decisions, and how a society based on direct democracy might look. This chapter provides a detailed guide for using consensus in your group. The tools described below are based on decades of experience in groups such as housing and workers' co-operatives. With commitment, they really do work and making decisions by consensus can be the bedrock of transforming our world and our relationships with each other.

what is consensus decision making?

Consensus is a decision-making process that works creatively to include all the people making the decision. Instead of simply voting for an item, and letting the majority of the group get their way, the group is committed to finding solutions that *everyone* can live with. This ensures that everyone's opinions, ideas and reservations are taken into account. But consensus is more than just a compromise. It is a process that can result in surprising and creative solutions – often better than the original suggestions. At the heart of consensus is a respectful dialogue between equals, helping groups to work together to meet both the individuals' and the group's needs. It's about how to work *with* each other rather than 'for' or 'against' each other.

Making decisions by consensus is based on trust and openness – this means learning to openly express both our desires (what we'd like to see happening), and our needs (what we have to see happen in order to be able to support a decision). If everyone is able to trust each other and talk openly, then the group will have the information it requires to take everyone's positions into account and to come up with a solution that everyone can support.

It may take time to learn how to distinguish between our desires and needs: after all most of us are more used to decision making where one wins and the other loses. In this kind of adversarial decision making we are often forced to take up a strategic position of presenting our desires as needs.

Box 4.1 *Guidelines for consensus building*

* Be respectful and trust each other. Don't be afraid to express your ideas and opinions.
* Don't assume that someone must win and someone must lose. Look for the most acceptable solution for everyone.
* Think before you speak, listen before you object. Listen to others' reactions, and consider them carefully before pressing your point.
* Remember that the ideal behind consensus is empowering not overpowering, agreement not majorities/minorities.

Conditions for good consensus

For good consensus building to be possible a few conditions need to be met:

Commitment to Reaching Consensus on all Decisions: Consensus requires commitment, patience, tolerance and a willingness to put the group first. It can be damaging if individuals secretly want to return to majority voting, just waiting for the chance to say 'I told you it wouldn't work.'	**Active Participation:** If we want a decision we all can agree on, we all need to play an active role in the decision-making.	**Clear Process:** Everyone needs to share an understanding of how consensus is being used. There are lots of variations of the consensus process, so even if people are experienced in using consensus they may use it differently to you! Explain the process at the beginning of the meeting.
	Good Facilitation: Appoint facilitators to help your large group meeting run more smoothly. The facilitators are there to ensure that the group works harmoniously, creatively and democratically.	**Trust and Respect:** We all need to trust that everyone shares our commitment, and respects our opinions and equal rights.
	Common Goal: Everyone at the meeting needs to be united in clear common goal — whether it's the desire to take action at a specific event, or a shared ethos. Being clear about the shared goal helps to keep a meeting focused and united.	

Figure 4.1   Conditions for good consensus

Source: Seeds for Change.

The consensus process

The dialogue that helps us to find common ground and respect our differences can take different formats. Some groups have developed detailed procedures; in other groups the process may be more organic. What process you use depends on the size of the group and how well people know each other. Below we outline a process for groups no larger than 15–20 people. Later on we discuss the spokescouncil process, which works for groups of hundreds, and even thousands, of people.

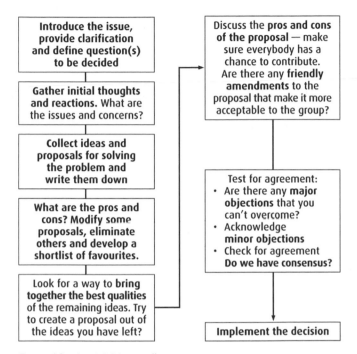

Introduce the issue, provide clarification and define question(s) to be decided

Gather initial thoughts and reactions. What are the issues and concerns?

Collect ideas and proposals for solving the problem and write them down

What are the pros and cons? Modify some proposals, eliminate others and develop a shortlist of favourites.

Look for a way to **bring together the best qualities** of the remaining ideas. Try to create a proposal out of the ideas you have left?

Discuss the **pros and cons of the proposal** — make sure everybody has a chance to contribute. Are there any **friendly amendments** to the proposal that make it more acceptable to the group?

Test for agreement:
• Are there any **major objections** that you can't overcome?
• Acknowledge **minor objections**
• Check for agreement **Do we have consensus?**

Implement the decision

Figure 4.2   A model for small group consensus

Source: Seeds for Change.

Dealing with disagreement in consensus

Consensus aims to reach a decision that everyone can live with. So what can be done when we need to reach agreement and we seem to be poles apart? To find a solution that works for everyone we have to understand the underlying problems that lead to the differing points of view and then come up with ways of addressing them: there are often specific problems causing the failure to reach agreement. These can often be dealt with by facilitation and are explored later in this chapter.

For those times when there is continued disagreement over a decision that needs to be taken, consider the following options:

★ The major objection (block or veto): Using your veto will stop the proposal going ahead, so think carefully before doing it. But don't be afraid to veto when it's relevant. A veto means: 'If this decision went ahead I could not be part of this project.' If someone expresses a major objection, the group discards the proposal and starts working on a new one. People often ask what happens if the rest of the group is unwilling to respect the veto. This is a difficult situation where the group needs to decide whether the proposal is so important to them that they will risk the person who objects leaving the group. The ideal is never to be in a situation where a major objection is being raised in the first place. The key to consensus building is to identify areas people feel strongly about early on in the process, so that any proposals already take them into account.

★ The minor objection (stand aside): There will be times when you want to object, but not veto. In those situations you can 'stand aside'. Standing aside registers your dissent, and says clearly that you won't help implement the proposal. A stand aside means: 'I personally can't do this, but I won't stop others from doing it.' The person standing aside is not responsible for the consequences, but also isn't stopping the group from going ahead with the decision.

★ Agree to disagree: The group decides that no agreement can be reached on this issue. Imagine what will happen in six months, a year or five years if you don't agree. Is the decision still so important?

★ The Fridge: Put the decision on ice, and come back to it in an hour, a day or a week. Quite often when people have had a chance to cool off and think it through things can look quite different.

★ Backup options: Some groups have fallback options when no agreement can be reached.

(a) Allow the person most concerned to make the decision.
(b) Put all the possibilities into a hat and pull one out. Agree in advance on this solution.
(c) Some groups have majority voting as a backup, often only after a second or third attempt at reaching consensus, and requiring an overwhelming majority such as 80 or 90 per cent.

★ Leaving the group: If one person continually finds him/herself at odds with the rest of the group, it may be time to think about the reasons for this. Is this really the right group to be in? A group may also ask a member to leave.

Facilitating the consensus process

Facilitation helps a group to have an efficient and inclusive meeting. Facilitators are essentially helpers. They look after the structure of the meeting, making sure everyone has an opportunity to contribute, and that decisions are reached.

Facilitation is a vital role that needs to be filled at every meeting. In small groups this function may be shared by everyone or rotated informally. Difficult meetings or meetings with a larger number of participants (more than eight or ten people) should always have clearly designated facilitators. However, all members of the meeting should always feel responsible for the progress of the meeting, and help the facilitator if necessary.

Box 4.2 A facilitator's skills and qualities

* Little emotional investment in the issues discussed. If this becomes difficult, step out of role and let someone else facilitate.
* Energy and attention for the job at hand.
* Understanding of tasks for the meeting as well as long-term goals of the group.
* Good listening skills including strategic questioning to be able to understand everyone's viewpoint properly.
* Confidence that good solutions will be found and consensus can be achieved.
* Assertiveness that is not overbearing – know when to intervene decisively and give some direction to the meeting.
* Respect for all participants and interest in what each individual has to offer.
* Clear thinking – observation of the whole group.
* Attend both to the content of the discussion and the process. How are people feeling?

Depending on the group a facilitator might:

☆ Help the group decide on a structure and process for the meeting and keep to it.

☆ Keep the meeting focused on one item at a time until decisions are reached.

☆ Regulate the flow of discussion – drawing out quiet people and limiting over-talking.

☆ Clarify and summarise points, test for consensus and formalise decisions.

☆ Help the group in dealing with conflicts.

Facilitation roles

One facilitator is rarely enough for a meeting. Depending on the size of the group and the length of the meeting some or all of the following roles may be used:

☆ The facilitator helps the group decide on and keep to the structure and process of the meeting. This means running through the agenda point by point, keeping the focus of the discussion on one item at a time, regulating the flow of the discussion and making sure everyone participates. The facilitator also clarifies and summarises points and tests for consensus.

☆ The co-facilitator provides support such as writing up ideas and proposals on a flip chart for all to see or watching out for rising tension, lack of focus, flagging energy. They can also step in and facilitate if the facilitator is flagging, or feels a need to take a position on an issue.

☆ Keeping a list of speakers and making sure they are called to speak in turn can either be taken on by the co-facilitator or it can be a separate role.

☆ The minute taker notes down proposals, decisions and action points for future reference. They also draw attention to incomplete decisions – for example who is going to contact so and so, and when.

☆ The timekeeper makes sure each agenda item gets enough time for discussion, and that the meeting finishes at the agreed time.

☆ The doorkeeper meets and greets people on the way into the meeting, checks that everyone knows what the meeting is for, and hands out any documents such as minutes from the last meeting. This makes new people feel welcome, and brings latecomers up to speed without interrupting the meeting.

Common problems and how to overcome them

These two examples show how important it is to get to the bottom of the underlying issues when things get tricky in a meeting. Develop your ability to spot problems, the underlying reasons for them, and how to deal with them. The more trust and understanding there is in a group the easier it will become to overcome problems. Facilitation can help supply the tools to avoid problems in the first place and help deal with them creatively if they do occur.

Box 4.3a   Problem 1

Tom, with lots of experience, confidence and a loud voice, is talking all the time and dominating the meeting. Hardly anyone else gets a chance to speak.

Underlying causes

* A lack of understanding of the consensus process on behalf of Tom coupled with an unwillingness of the rest of the group to challenge his behaviour.

Possible solutions

The facilitator can equalise speaking time by using tools such as:

* Introduce a go-round – each person speaks in turn for a set amount of time.
* At the beginning of the meeting set a limit on how many times a person can speak.
* Proactively ask other people for their opinion: 'Thank you, Tom, for your great ideas. What do other people think?'

Box 4.3b
Problem 2

People are coming up with lots of ideas, but the discussion is going nowhere. People keep going off on tangents.

Underlying causes

* Lack of structure and focus for the discussion.
* Weak facilitation.

Possible solutions

The discussion can be moved on from its creative phase to making decisions:

* Write all ideas up on a flip chart for all to see.
* Discuss one idea at a time, recording pros and cons for each one.
* When people bring up tangential issues, record them for discussion later. Avoid getting sidetracked.
* Check if the facilitator needs a break or support.

tools for meetings

Here is a selection of tools you can use at various stages of a meeting to make it efficient and enjoyable for all. It is always a good idea to explain to people what tools you are using and why.

(a)   At the beginning of the meeting

   ☆ Consensus training: Running pre-meeting 'introduction to consensus' sessions can make meetings more inclusive for everyone, and avoid conflict that arises from a misunderstanding of the process.

   ☆ Setting up the meeting venue: It's important that the space, and the way you use the space, doesn't isolate or alienate anybody. Is everyone able to hear and see clearly? Some rooms have very bad acoustics that require people to shout to be heard. Others have fixed seating or columns that restrict people's view and their ability to participate. Is the venue accessible to everyone?

   ☆ Group agreements and ground rules: Agree at the beginning of the meeting on how the meeting will be run. This prevents a lot of problems from occurring in the first place. It also makes it easier for the facilitators to challenge disruptive behaviour, as they can refer back to 'what we all agreed'. Possible ground rules might include: using consensus, hand signals, not interrupting each other, active participation, challenging oppressive behaviour, respecting opinions, sticking to agreed time limits, and switching off mobile phones.

   ☆ Clear agendas: These can help make a meeting flow more easily. Sort out the agenda at the start of the meeting or even, with the participation of the group, in advance. Be realistic about what can be achieved in the time you've got, and decide which items can be dealt with at a later meeting. Set time limits on each agenda item to help the meeting end on time. Make sure that everyone has an up to date copy of the agenda or write it up on a flip chart for everyone to see.

   ☆ Using hand signals: These can help meetings run more smoothly and helps the facilitators spot emerging agreements. It is important to explain what hand signals you will be using *at the start of the meeting* to avoid confusion!

(b)   When making a decision
        Not every tool is suitable for every stage of the consensus process. Think carefully about when you would use which tool and why.

★ Go-rounds: Everyone takes a turn to speak without interruption or comment from other people. Go-rounds help to gather opinions, feelings and ideas as well as slow down the discussion and improve listening. Make sure that everyone gets a chance to speak.

★ Idea storm: Ask people to call out all their ideas as fast as possible – without censoring them. All ideas are welcome – the crazier the better and helps people to be inspired by each other. Have one or two note takers to write all ideas down where everyone can see them. Make sure there is no

I want to contribute to the discussion
Raise a hand or forefinger when you wish to contribute to the discussion.

Technical point
Make a T-shape with your hands to indicate a proposal about the process of the discussion, eg. 'let's have a break'.

'I agree' or 'Sounds good!'
Silent Hand clapping. Wave your hands with your fingers pointing upwards to indicate your agreement. This gives a very helpful visual overview of what people think. It also saves time as it avoids everyone having to say 'I'd just like to add that I agree with...'.

Figure 4.3   Consensus hand signals

Source: Seeds for Change.

discussion or comment on others' ideas at this stage. Structured thinking and organising can come afterwards.

★ Show of hands or straw poll: An obvious but effective way of prioritising items or gauging group opinion. Make sure people understand this is not voting, but to help the facilitators spot emerging agreements.

★ Clear process: Used when dealing with multiple proposals. For example, if you plan to consider ideas in turn, let people know they'll all be considered and given equal time. Otherwise some people may well be unco-operative because they can't clearly see that there is time set aside to talk about their idea and may feel like they're being ignored. If you're putting some ideas to one side, after a prioritisation exercise for example, you might like to ensure their 'owners' have agreed and understand the reasons why.

★ Pros and cons: Got several ideas and can't decide which one to go for? Simply list the benefits and drawbacks of each idea and compare the results. This can be done in a full group, in pairs or small groups, working on the pros and cons of one option and reporting back to the group.

★ 'Plus-Minus-Implications': A variation of the simple 'pros and cons' technique. It will help you decide between a number of options by examining them one by one. Create a simple table with three columns titled Plus, Minus, and Implications, and write 'positives', 'negatives' and 'implications' in each.

★ Breaks: Taking a break can revitalise a meeting, reduce tension, and give people time to reflect on proposals and decisions. Plan a 15-minute break at least every two hours and take spontaneous breaks if the meeting gets too heated or attention is flagging.

(c)   At the end of a meeting

★ Evaluation and constructive feedback: Evaluation allows us to learn from our experiences. It should be a regular part of our meetings and workshops as it gives us the chance for honest feedback on the process and content of the event, allowing us to improve in the future. Everyone who participates in an event should be encouraged to take part in its evaluation.

consensus in large groups – the spokescouncil

When making decisions in a large group there is a tendency to have one large meeting with hundreds of people. One of the problems with this format is that the large majority of people do not have a chance to speak due to time constraints. Instead it is usually dominated by a few confident people. This is not a good starting point for reaching consensus, which depends on mutual understanding and trust. Good consensus building is based on working in small groups where everyone contributes to the discussion.

The spokescouncil was developed to address this problem. It enables large numbers of people to work together as democratically as possible, allowing the maximum number of opinions and ideas to be heard in an efficient way. Many groups such as social centres and large workers' co-operatives use this process successfully as well as peace, anti-nuclear and environmental movements around the world.

How a spokescouncil works

In a spokescouncil the meeting breaks up into smaller groups to enable everyone to express their views and take part in discussions. Small groups can be either based on working groups, in regional groupings based on shared political analysis, or be entirely random. People in each small group discuss the issues and come up with proposals and concerns.

Each group sends a delegate (or *spoke*) to the spokescouncil meeting, where all the spokes present the proposals and concerns of their group. The spokes then come up with proposals that they think might be acceptable to everyone and check back with their groups before a decision is taken.

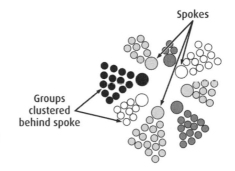

Figure 4.4   A typical spokescouncil

Source: Seeds for Change.

For a spokescouncil to work effectively the role of the spoke needs to be clearly defined. A group can choose to use the spoke as a voice – feeding back to the group the collective, agreed thoughts. Or the small group might empower their spoke to make certain decisions based on their knowledge of the small group. Being the spoke is not easy – it carries significant responsibility. You might like to rotate the role from meeting to meeting or agenda item to agenda item. It also helps to have two spokes, one presenting the viewpoints and proposals of their small group, the other to take notes of what other groups have to say. This helps to ensure that ideas don't get lost or misrepresented in the transmission between small groups and the spokescouncil. Spokescouncils require good facilitation by a team of at least three facilitators, which work well together and who are skilled at synthesising proposals.

This process works regardless of whether everyone involved is in the same location or geographically dispersed. Where small groups are based in different places, the spokescouncil either involves a lot of travel for the spokes or the spokes communicate via telephone conferences and chat rooms.

If all the people involved in making the decision are together in the same place, it works well if groups sit in a cluster behind their spoke during the spokescouncil. Groups can hear what is being discussed and give immediate feedback to their spoke. This can make the spokescouncil more accountable and reduce the need for repeating information.

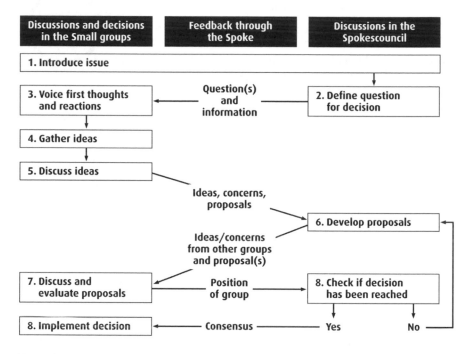

Figure 4.5   A model for spokescouncil consensus

Source: Seeds for Change.

Variations

If the issue impacts strongly on the needs of the people involved, then an additional step can be built in where small groups give information on their particular needs via the spokescouncil before starting to gather ideas. When there are just a few people with strongly opposing views that seemingly can't be resolved within the format of the spokescouncil we have successfully used the 'back of the barn' technique. This involves those with strong views having a separate meeting with the aim of working out a proposal that they can all agree to. This definitely benefits from an experienced facilitator who can help people express and listen to each other's concerns and needs.

Making consensus work with thousands of people

The spokescouncil itself is limited by the number of spokes that can have a meaningful exchange of information and discussion in the spokescouncil. In our experience a spokescouncil becomes much more difficult when more than 20 small groups are represented. If the maximum size of each small group is 20 people as well, this gives a natural limit of about 400 people for which the spokescouncil works.

To make consensus decision making possible with thousands of people, peace and anti-nuclear movements have developed a three-tier system, where small groups are affiliated in clusters who then send spokes to an overall spokescouncil.

The key to making this work is to make decisions at the most local level possible. Not every decision needs to be taken by everyone. The spokescouncil should be reserved for only the most important decisions, generally at a policy level. It is often the facilitators that will spot proposals that do not need to be decided in the whole group. For example, discussion around the wording of a press release should take place in the small working group that is actually writing it. This group can consult with everyone else for their ideas and preferences, but this is different from attempting to reach a decision with everyone. Consensus is based on trust and good will, even more so in a large group.

conclusion

Consensus is about participation and equalising power. It can also be a very powerful process for building communities and empowering individuals. Despite sometimes taking longer to achieve, consensus can actually save time and stress, because the group doesn't have to keep revisiting past decisions – they were fully supported at the time they were made. Don't be discouraged if the going gets tough. For most of us consensus is a completely new way of negotiating and making decisions – it takes time to unlearn the patterns of behaviour we have been brought up to accept as the norm. Consensus gets much easier with practice, and its true potential is often only recognised after a difficult decision has been reached in a way that everyone is happy with.

Seeds for Change are a UK based collective of activist trainers providing training for grassroots campaign groups. They also develop resources on consensus, facilitation and taking action, all of which are available on their website www.seedsforchange.org.uk

resources

Books

Try your local library first – they are generally quite happy to order or even buy books for you. If you decide to buy a book, get it from one of the radical/independent bookshops – they all do mail order! In the UK try News from Nowhere in Liverpool (0151 708 7270) or Housmans Bookshop in London (020 7278 4474).

Albert, Michael (2006). *Realizing Hope. Life Beyond Capitalism*. London: Zed Books.

Beer, J. and E. Stief (1997). *The Mediator's Handbook.* 3rd edn. Gabriola Island, BC: New Society Publishers. (Developed by Friends Conflict Resolution Programs.)

Coover, V., E. Deacon, C. Esser and C. Moore (1981). *Resource Manual for a Living Revolution.* Gabriola Island, BC: New Society. (A complete manual for developing your group. Sadly out of print now, but it's worth trying to get your hands on a copy!)

CrimethInc. (2000). *Days of War Nights of Love – CrimethInc for Beginners; Demon Box Collective.* (A book of thoughts, inspirations and texts – read this if you want to think more about the issues raised in this chapter.)

CrimethInc. (2005). *Recipes for Disaster. An Anarchist Cookbook.* Montreal.

Fisher, S. et al. (2000). *Working with Conflict.* London: Zed Books.

Freeman, J. (1972). *The Tyranny of Structurelessness*, www.struggle.ws/anarchism/pdf/booklets/structurelessness.html (Pamphlet about informal hierarchies in small groups.)

Gastil, J. (1993). *Democracy in Small Groups – Participation, Decision Making and Communication.* Philadelphia: New Society. (Looks at various types of decision making processes.)

Gordon, N. and P. Chatterton (2004). *Taking Back Control – A Journey through Argentina's Popular Uprising.* Leeds: School of Geography, University of Leeds. (Eyewitness accounts of the developing parallel institutions in Argentina.)

Kaner, S., L. Lind, C. Toldi, S. Fisk and D. Berger (1996). *Facilitator's Guide to Participatory Decision-Making.* Gabriola Island, BC: New Society Publishers.

Potter, B. (1996). *From Conflict to Cooperation – How to Mediate a Dispute.* Berkely, CA: Ronin Publishing.

Starhawk (2002). *Webs of Power: Notes from the Global Uprising.* Gabriola Island, BC: New Society Publishing.

Ward, C. (1988). *Anarchy in Action.* London: Freedom Press. (Many of the ideas in these chapters have been developed and used extensively by anarchists. This is one of many books providing an introduction.)

Werkstatt für Gewaltfreie Aktion Baden (2004). *Konsens – Handbuch zur Gewaltfreien Entscheidungsfindung.* Gewaltfrei: Leben Leben. (Probably the most current and comprehensive book on consensus decision making – includes exercises, detailed descriptions and exercises; in German; buero.karlsruhe@wfga.de.)

Websites

Blatant Incitement Project www.eco-action.org/blinc
Groundswell www.groundswell.org.uk
Participatory Budgeting www.participatorybudgeting.org.uk
Rant Collective www.rantcollective.net
Seeds for Change www.seedsforchange.org.uk
Skillsharing www.skillsharing.org.uk

5 why society is making us sick

Tash Gordon and Becs Griffiths

your health, whose responsibility?

Your health. Is it something you take for granted, feel helpless to improve or something you feel you understand and take responsibility for? Our health is determined by many factors and we want to explore these and talk about how we can affect them. Crucial to this is how our society functions. While we agree that our industrialised capitalist societies have brought some enormous improvements in health, we also need to look at how they constrain it and how a centralised and corporate influenced medical profession has reduced our ability to manage our own health. This chapter is about how we need to change society and our relations with others in order to improve our health. As we discuss, societies that are the most equal, not the wealthiest, are the healthiest. It is possible to act for ourselves to improve our health and that of others, and this chapter and the next provide some pointers.

What guarantees our health is complex and difficult to define. We have chosen a broad view of health which encompasses our physical, mental and social well-being. Good health in all these areas is the basis of our lives and therefore of paramount importance. Illness undermines our ability to live as we wish and our ability to challenge the conditions that make us sick. Our immediate environment is the major factor determining our health. Put simply, genetic make-up dictates a number of rare conditions and a predisposition to many more common diseases, but the world around us plays a far larger role. Health is affected by material conditions, including where we work and what we eat, but there are also, largely ignored, social conditions: how vulnerable or oppressed we are, who we have to support us, the amount of conflict we face, the fear we live with, how powerless or empowered we feel, the amount of control we have in our lives, and how safe we are from violence. The greater our control over our own lives, the more opportunities there are to improve our health.

The motivation and interest for us to write this chapter comes from both our personal and professional experiences. One of the authors (Tash Gordon) is an inner-city GP. Daily she confronts the impact society has on our health and is acutely aware of the way in which the personal situations of patients limit the action we can take to improve our health. A recent article in the GPs' (family doctors') journal, the *British Journal of General Practice*, suggested GPs should stop being involved in social work issues like child abuse, homelessness and poverty, and stick to treating illness. This attitude results in a narrow definition of ill health; one far from the reality of the author's work where she might encourage a depressed patient to challenge the racial harassment that undermines her or for the author to write a letter supporting a patient's request for appropriate housing. The author's experience of working with autonomous health systems in Chiapas and Argentina taught her that the more involved people are in their health the more responsibility and action they take.

The other author (Becs Griffiths) is part of a feminist health collective and is studying herbal medicine. She has participated in many workshops around women's health, reclaiming knowledge of our anatomies and menstrual cycles, pregnancy, abortion, breast health and more. She is also studying herbal medicine at university and as part of her studies she is involved in a herbal clinic. She has been consistently shocked by how people are treated by doctors, how little they are listened to, how little information doctors want to know about their lives, how many drugs they have been put on, how many drugs they take without knowing why, and how little people know about their own bodies. These experiences have politicised her and made her realise how important it is to start learning and educating each other about our bodies and how to treat ourselves so we are less reliant on 'Western' medicine that treats the symptoms, not the causes.

society is sick

If we want to create a healthier society we have to challenge the root causes of illness and disease. Acknowledging the type of society that we live in and how it impacts on our lives is an important start. We live in a society that is largely industrial, urban, based on hierarchy and capitalist in orientation, that is to say, an economic system that is reliant on the exploitation of people and natural resources for the accumulative wealth of a few. This has resulted in a society that prioritises profits and maintains power in the hands of a few, with little consideration of the quality of people's lives. This type of society impacts on our lives in many ways: the pressure to work, the type and amount of work we have to do, how our time is valued, family structures, education systems, the quality of food we eat, pollution, alienation, poor housing and, of course, the medical system available to us.

This has different effects in different countries but ultimately most people are reliant on selling their time so that they can earn enough money to live. The negative health impacts vary in different parts of the world. In the global North, social isolation is detrimental to mental health (seen through the increasing number of suicide attempts by younger and younger people, and those suffering from depression and anxiety), whilst increasingly sedentary lifestyles, the relatively high cost of healthy food, and the availability of energy dense food is creating an obesity epidemic. The global South is used as a resource to keep the North rich, which undermines sustainable communities and traditional ways of life, replacing them with exploitative jobs, poor housing and loss of control over their lives. The control from the global North is increased through free trade agreements, structural adjustment programmes and military power, while war and the negative effects of climate change are a daily reality for many. All of these aspects of our globalised society are detrimental to health. As always, the poorest are most affected.

Our health depends not just on income but on a complex set of hierarchies, including wealth, race, class, gender, disability, age, sexuality and cultural background. Being black in the global North, for example, means you are more likely to be detained for compulsory psychiatric treatment and stay in hospital longer than white people and also more likely to be prescribed drugs or electroconvulsive therapy (ECT) rather than psychotherapy or counselling. If your sexuality challenges society's norms you will find that health professionals often have no training in issues specific to same sex relationships. As a woman you have a one in four lifetime chance of suffering domestic violence and its psychological and physical effects and you may find that people accept this as a normal part of a relationship. Class is a major factor: men in central Glasgow have an average life expectancy of 69.9 years, while it is 86.2 years for women in the Royal Borough of Kensington and Chelsea in London. Poor urban areas with high minority group populations within the USA have life expectancies and infant mortalities similar to those in Guatemala. Our position in these hierarchies affects our health through access to health care, education, good diet, housing and other opportunities that cumulatively mean poorer health and less time, energy or belief that this can change.

Living in a complex, urban, capitalist and hierarchical society in which many individuals and institutions have control over our lives creates stress. There is a proven link between levels of stress and the position we occupy in a social hierarchy – the lower the position, the higher the stress. The accumulative effect of this stress is chronic anxiety, which has major impacts on health. This means that illness is not only caused by specific material situations, such as poor diet or damp housing, but can also be triggered by the anxiety caused by what we feel and think about our material and social circumstances (Wilkinson 2001). The alienation that develops because of hierarchies can impact on our health in terms of loneliness, depression and anxiety. All this not only leads to conditions such as high blood pressure, heart problems and lowered immunity, but also patterns of destructive behaviours like alcoholism, other addictions, violence and self-abuse. Similarly, studies have shown that it is not the richest of the developed countries that have the highest life expectancies, but those that have the most egalitarian distribution of income and less pronounced social hierarchies – an indication that if society is more equal people are more healthy (Wilkinson 1992).

The loss of traditional medical knowledges

From the fifteenth century in Western Europe there was a clear break away from the tradition of lay healers and communal medicine. This was mainly caused by the witch hunts that killed hundreds of thousands of women and lay healers over the next two centuries, resulting in the effective loss of their knowledge (Frederici 2004). The *Malleus Maleficarum*, written in 1484, was one of the first and most influential witch finder manuals, including specific torture techniques, and sparked a wave of manuals and witch trials. The church and the state in the form of the legal system and medical establishment had the power to declare who actually was a 'witch', and who carried out the trials and executions.

The suppression of lay healing and the rise of male professionals was not a natural process but an active takeover. The witch hunts specifically targeted female healers and especially midwives and those offering help with contraception. 'No-one does more harm to the Catholic Church than midwives' stated the *Malleus Maleficarum*. The Papal Bull of 1484 declares 'witches destroy the offspring of women... They hinder men from generating and women from conceiving.' In 1548 Reginald Scott wrote: 'At this day it is indifferent to say in the English tongue, "she is a witch or she is a wise woman"' (Thomas 1971). Many of the trials featured the topic of health. For example, if a woman cured someone, and that person became ill again, it was seen as witchcraft, and therefore a crime. The male healers, wise men and magicians were not killed and often ended up in positions of equal standing with the new physicians.

Historians have various explanations for this attack on women healers, which include physicians wiping out competition or covering up for incompetence or unexplained deaths – for example, they had little knowledge of cancer or strokes. A deeper and more political explanation is a contestation over control of the body, the female body in particular, and especially reproduction. This was due to a desire to control birth rates (adversely affected by women's knowledge of contraception and abortion) and more fundamentally to distance people from control over their bodies in parallel with new forms of discipline associated with capitalist work appearing over this same period (Frederici 2004).

After this bloody campaign of terror over two centuries had succeeded in effectively wiping out the networks of female lay healers, the poor population were faced with a choice between expensive, and largely unaffordable, trained physicians, magicians more specialised in finding spells than biology, or the demoted, untrained and unskilled neighbour. Women's ability to control their reproduction was hugely diminished as was the tradition of community based knowledge.

Western 'medicine' as we know it became established over this same period, with many of the men praised as the fathers of modern science deeply involved in the witch hunts. For example, Francis Bacon exposed the evil of witches alongside his more famous 'scientific rationality'. This new 'age of reason' included a ferocious attack on women's bodies (Ehrenreich and English 1973). Medicine grew in prestige as a secular science and a profession. By the fourteenth century there were licensing laws that prohibited all but university trained (and therefore male) doctors to practice, thereby providing a legal framework which complemented the physical persecution of lay healers.

As the training of male professionals developed and became the norm in the modern period, it ignored lay healers' knowledge built up over centuries which used empirical methods and had developed an extensive understanding of bones, muscles, herbs and drugs. As the largely male medical profession became dominant and their views were seen to be those of the 'expert', people were forced to become consumers of this type of medicine due to lack of alternatives. No attempt was made to share new modern medical knowledge such as anatomy and physiology, by the medical profession, and this discouraged people from developing an understanding of their own health and taking responsibility for it. As with many other aspects of the modern period, the irony was that despite massive progress in human understanding, this was limited to a small elite and actually fostered dependency and hierarchies.

However, it is fair to say that in many parts of the world, such as China or Africa, lay knowledge and practitioners faced less persecution and retained a more important role. Some of this has become part of Western medicine through the growing use of complementary therapies. Also the industrial era, while bringing many negative effects on health, also saw dramatic improvements in the management of sanitation and public health. Overall though, understanding the political construction of this type of Western scientific medicine enables us to question expertise, reclaim knowledge of lay medicine and therefore begin to reconstruct a different type of health care.

Medicine – healing or creating disease?

Over the last century the medical profession has played a greater role in everyday life, for example, in pain, sickness, pregnancy, childbirth, menopause and death. Fortunately medical training has improved over the years and doctors are both well trained in the health sciences and learn through real experiences with patients. This goes hand in hand with a welcome increase in evidence based treatments for a variety of conditions from meningitis to spinal injury. But with this, normal bodily processes have been brought under professional medical scrutiny and control. Such med- icalisation of life is disempowering as it leads to a loss of knowledge and resources to care for oneself rather than sharing knowledge about how to deal with illness. For example, over the last 200 years childbirth has become increasingly medicalised with high levels of caesarean sections and other medical interventions to speed up labour or give pain relief – actually making it harder for women to give birth (Vincent 2002). Births in hospitals are seen as the norm compared to home births despite no evidence of this being safer and plenty of evidence that women have less choice in their care (Johanson et al. 2002).

More medical interventions can have negative impacts on health. Taking the example of the side effects of medications, in the UK over 6800 people die per year from adverse drug reactions and 6.5 per cent of all hospital admissions are due to these reactions. Menopause, which is a natural and inevitable event, became seen as a disease of oestrogen deficiency to be treated with hormone replacement. Unfortunately, it took several years to produce evidence of its detrimental effects, such as increased rates of breast cancer. All medication has risks and benefits and clearly, while serious illnesses need effective treatment, the risk-benefit ratio must be carefully considered – for example, are hot flushes problematic enough to justify the increased risk of breast cancer? Taking a pill may seem the simple solution for depression but non-pharmaceutical options like increasing exercise and support in mild depression may be safer and more appropriate.

Our interaction with our health

We often only become aware of our health when we have a problem and then the path we take is to look to the medical profession for an answer. This can be problematic for a number of reasons.

First, although the majority of health experiences we have can be self-managed, with a small minority of illnesses a well trained health professional is desirable. There is nothing wrong with people accumulating large amounts of health knowledge, be they doctors, herbalists, witches, traditional healers. This is useful and necessary and has happened throughout history. Unfortunately doctors have been elevated to a position where their knowledge is respected the most – reflected in their social and economic position in society compared to any other health worker. This creates an imbalance in power which is often seen in the way doctors relate to their patients, as well as to other health workers.

Second, doctors are reluctant to investigate the social causes of illness. Within the health profession as a whole, there is insufficient examination of the political causes of ill health. Nor are the power relations acknowledged that are embedded in gender, class, race, caste, sexuality, age, ability, and their cumulative effect on people's bodies and lives. These power relations become institutionalised and are indeed replicated within the medical system. Research indicates that working-class women and black women are likely to receive less favourable treatment in hospitals than other social groups (Douglas 1992).

Third, there are general problems in relying on any 'professional' as their existence can discourage or prevent others from taking responsibility for their health. It is important to be able to challenge 'experts', question their evidence, motivations and judgements, and their influence over our lives and health care to ensure it is appropriate for us. In many health consultations there is an imbalance of power with people being told what to do with little discussion about whether they actually understand or not.

There is a large body of evidence that shows that health improves when there is a shared understanding of the problem and that the person and the health professional manage it together. Medicine is an art and finding the correct treatment requires an in-depth understanding of the individual. To ensure shared understanding the health professional must respect the person's autonomy (the most important medical ethical principle) and the patient must play an active role in their own care.

Many people, through negative health experiences, have become positively politicised. Unsatisfactory care might result in the pursuit of alternatives to mainstream health care methods or to questioning the causes of a health problem, leading to a better understanding of the power structures in society. People have repeatedly taken action to try and claim more control over factors affecting their health, realising that direct, autonomous input is often necessary to improve a detrimental social situation.

More disease = more profit

There are important criticisms to be levelled at the close ties between the medical profession and the pharmaceutical industry. Whose interest do they prioritise – that of the patient or the drug company? More money is invested in research and development of medications to be marketed in the global North where drugs can be sold at higher cost to more people. To increase market size pharmaceutical companies have been involved in changing and creating new definitions of diseases and then promoting these to both the medical profession and potential users of their treatments.

Everyday life occurrences are turned into medical problems, mild symptoms are portrayed as serious, and risks become diseases. This can make people obsessed with their health and disguise the social or political causes of health problems. These campaigns are normally linked with a drug company promoting a new product for the 'disease'. Recent examples of this include baldness, irritable bowel syndrome, social phobia, risk of osteoporosis and erectile dysfunction (Moynihan et al. 2002). No doubt these drugs tap into the very real desires that many of us have to be, for example, less shy or bald or fat, or more virile. They make us feel these are medical problems with easy pharmaceutical answers rather than looking at the social context or origins. Of concern is the way drug companies use these legitimate concerns to sell drugs and make easy money, which distracts us from looking at deeper causes.

The consumer marketing of obesity drugs often lacks reference to lifestyle changes that are needed for a healthier lifestyle, and certainly lacks any reference to the responsibility of food manufacturers to reduce the salt, sugar and fat content of their products.

Drug companies are able to exert their influence in a number of ways. They are involved in committees that write the guidelines for how conditions are treated, such as blood pressure where they have influenced the decision to lower the level at which blood pressure needs treatment with medication. They are involved in the education of the medical profession – sponsoring many postgraduate educational events. They also use the media and patient groups to turn people into patients who need treating with their drugs (Moynihan et al. 2002).

Drug companies are patenting medicinal plants, which means people who have traditionally used them for centuries will have to pay to grow and use these plants. In many cases, the drug companies have spent time in traditional communities and used their information developed over years of shared community health to identify plants and their effects. Indigenous communities often only discover that it is possible to own the right to grow and use a plant when they are informed a patent exists and they can no longer do so. An example of this is the Neem tree that has been used for centuries in India as a medicine, toothpaste and fuel. US and Japanese companies have patented many of its constituents and the seeds are now only available at a very high price. Indian farmers are challenging the rights of these multinationals to the 'intellectual property rights' of the Neem tree.

Responding to the needs of shareholders and accountants, drug companies' main concern is profit. Birth control is a well documented example of how profit is put before human safety. Women have been made infertile, or have died in some cases, by the aggressive marketing and selling of unsafe products like the Dalkon Shield coil (Hartmann 1995). Drug trials are often carried out in the global South where no ethical approval is required and the subjects are often unaware that they are part of a trial, or that the drug they are receiving is not licensed. Much of this has come into our consciousness through dramatic portrayal, such as the recent film *The Constant Gardener* and a Channel Four documentary in the UK on drug testing in India (Iheanacho 2006).

Drug companies are not the only multinationals that profit at the expense of our health. Nestlé do so with their sale of breast milk substitutes in areas where drinking water is unsafe. The World Health Organisation (WHO) estimates that 1.5 million infants die around the world every year because they are not breastfed. Where water is unsafe, a bottle-fed child is up to 25 times more likely to die as a result of diarrhoea than a breastfed child. Nestlé are still cynically promoting milk substitutes for breast milk even though their aggressive baby food marketing practices have been exposed. Lifestyle health is a big earner and more and more high street shops, such as Boots, Tesco and Wal-Mart, are offering free health checks and health advice. Sure enough, they have a solution and a product for sale too.

Health spending

Increasing health care resources are often sold as the solution to our health problems but there is no definitive evidence that this is the answer. Funding research to increase understanding and develop new treatments of illnesses like cancer will undoubtedly improve the health of those suffering from these illnesses. But spending that targets the root causes which are often social may have more impact. Take, for example, the fact that those in lower social classes are more likely to develop heart disease or lung cancer. The major factors that have improved health in the developed world in the last century are improvements in sanitation and housing rather then increased spending on health. There are still large areas of the world without access to clean water or reasonable housing. Almost 50 per cent of all people in poor countries at a given time have a health problem caused by lack of water and sanitation, but this is not where the cash is going (UNDP 2006). Health policy should be based on an evaluation of where increased resources would make the most difference. The most effective use of resources would be in the global South, preventing and treating infective diseases. Even improving support for isolated older people in the UK with chronic disease is a priority over checking healthy people's cholesterol. But it's the latter that makes most political capital and cash for the pharmaceutical companies (Heath 2005).

Interestingly those living in richer countries with more access to health care have higher rates of self-reported illness than those living in poorer countries. One reason for this is that due to the emphasis on preventative health, we are assessed for diseases we may develop in the future and given medications to reduce risks, such as raised cholesterol. In theory by reducing all our risk factors for ill health we should improve our health. The reality is that a combination of thinking we may get ill in the future from heart attacks or strokes and the side effects of the medications negatively impact on our mental and physical health (Heath 2005).

There have been several reports on widening health inequalities in the UK and across the globe which suggest that these can only be addressed by tackling income inequality and social determinants of health (Kawachi and Kennedy 1997). Solutions provided by medication are easier than challenging or changing our society – but this is the only action that will bring improvements in health for all.

autonomous health

The combination of capitalism, medicalisation and corporatisation are detrimental to our health and we believe can only be tackled by taking an autonomous approach to our health based on self-management and self-empowerment. This requires an awareness that our health is in a constant state of flux affected by us and the world around us, some of which we can control and some which we can't. In difficult social situations it may be impossible to either have this awareness or to take any action to improve things – which is why real autonomy is only possible with radical social change. Without challenging capitalism and its widely pervasive influences we will only have limited control and choices. The structure of our society alienates and isolates us so we do not recognise that our problems, and therefore solutions, should be shared.

In taking a more autonomous, self-managed, approach to our health, our main point as outlined earlier is that everything about how we live and how our society functions impacts on our health. Taking more individual responsibility for our personal health is a positive progression, whether it is doing more regular exercise, finding emotional support or, more broadly, aiming to be better connected with our bodily experiences. However, individual self-improvement is insufficient as an end point to good health since collective action is needed to tackle the widespread levels of unhealthiness that characterise our society. Our consumption based society sells us a solution to our health problems, in the guise of the 'new age movement', by encouraging individual self-improvement – as long as you can afford it – in the form of vitamins, organic food, gym membership or massage. These may be good for an individual's health, but a substantial impact on health will only come with social change in which the lives of a much larger group of people are improved.

Autonomous health, to us, means a grassroots self-help approach relying greatly on preventative measures, as well as developing a political consciousness around the root causes of ill health. There needs to be a multifaceted approach to autonomous health involving different concepts and practical solutions including: self-help techniques; challenging the role of the 'expert' and demystifying medicine, confronting the causes of ill health on an individual and collective level; building support networks; acquiring skills and skill sharing; emphasising education and prevention; and taking action against factors which work against our health.

Figure 5.1   Zapatista health clinic – 'For Autonomous Health'

Source: Tash Gordon.

Aspects of autonomous health

Within the constraints of our society there are self-organising, grassroots political struggles and groups that have been inspiring and made significant differences to peoples' lives. But we cannot create spaces which totally escape the society we live in – we can't be completely autonomous. Collective organising can improve our health. It can be empowering, create new bonds and offer new answers, but it will not get rid of health inequalities or change some of the fundamentally damaging things about this world. Having control over our lives in a number of ways can give us increased control of health and hopefully improve it, and it can stop us feeling paralysed by feelings of powerlessness.

Box 5.1 The health system of the Movement of Unemployed Workers, Argentina

In Argentina the MTD (Movement of Unemployed Workers) block the country's road infrastructure to paralyse the government and negotiate regular payments to the unemployed people. Some of the money goes directly to individuals, but most goes to funding community bakeries, workshops, nurseries, libraries, cafes and pharmacies to meet basic needs of the people involved in the movement. The MTD is in the process of setting up an autonomous health system. They have started by asking each neighbourhood group to pick a few health workers who then form the health commission. The first task undertaken by the health workers is to ask each of their neighbourhood groups to define what they perceive health to be, the barriers to their health, what is lacking in the health care they receive and what health care they would like to have available. The movement's ongoing popular education, apart from defining the health commission's work, is trying to raise consciousness about problems the people face, the underlying causes of these and how people could take control to improve their own situation. The plan is to enable sympathetic health professionals to give basic health training. There are already some workers trained to dispense medications, and others will soon be receiving first aid training, addiction training and a medicinal plant course.

Self-help

Self-help is a term that became widely used during the second wave of the 1970s feminist movement. It encourages autonomy through information sharing, directly challenging the notion of the medical expert. Women formed small self-help groups to talk about their experiences of their bodies and of the medical establishment. By trusting their own subjective experiences they could explore their commonalities and their differences. They were able to develop political theories around gender and a political consciousness to tackle the cause of their oppression. Forming self-help groups can be a very important tool of autonomous health. It can create spaces where we support each other in processing the information we receive from doctors, the state and pharmaceutical companies, to explore for ourselves the problems and the solutions. It is important to realise that we are able to change some of the factors that cause health problems and this is much more effective and empowering than taking a tablet. There are many self-help groups now organised around every condition thinkable – a useful support network and way of sharing knowledge. The critique of this is that although useful, they are looking at each condition as a single issue without placing them in the broader context of how society impacts on our health.

Developing a political consciousness

We must realise that things will not change without taking on the politics that create the situations that are detrimental to our well-being. It is essential to understand how ill health is caused by the political system that we live in. For the Black Panther movement in the USA it was part of a wider movement to develop political awareness – as Blacks in the USA, they were oppressed and therefore education or health care provided by their oppressors would never meet their needs. Poor education, cheap alcohol and drugs enabled the white state to maintain the status quo.

Skill sharing

Sharing skills is an essential way of passing on knowledge, without relying on experts or paid officials, and building capacity and understanding that can be used for social change. Sharing skills and knowledge has been a long tradition in health. In many countries, for example, a number of activist or street medic collectives provide medical support during actions and demonstrations (see Chapter 6). Volunteers from US street medic collectives have also volunteered at the inspiring Common Ground Health Clinic. This was set up in Algiers, a poor area of New Orleans, following Hurricane Katrina. A wide range of medics and therapists staff the clinic including doctors, nurses, mental health counsellors, massage therapists, acupuncturists, osteopaths, physiotherapists, social workers and pharmacists. As well as running a clinic, they go

Box 5.2   The Sambhavna Clinic, India

In 1984 methyl isocyanate, a deadly toxic gas, leaked into the town
of Bhopal killing 20,000 people with an estimated 150,000 survivors
and tens of thousands of children born with deformities. The
company involved, Union Carbide, now owned by Dow Chemicals, paid
very little compensation, failed to clear up properly, and
allowed the groundwater to become contaminated. The
state health care has been inadequate with indiscriminate
prescriptions that have compounded the damage caused
by the gas exposure. Survivors, doctors, social workers
and activists set up Sambhavna Clinic in 1995. Its task
was to try and explore all the possible ways of improving the health
of the survivors. It is run with no formal hierarchy of jobs, with
every member of staff free to give opinions on every aspect of the
clinic's work. The co-ordinator's role is on a two-month rotational
basis so that everyone has to take responsibility for themselves and
everyone else. Decisions are made at weekly meetings by consensus.
    A study carried out by the clinic in 1996 showed that a dozen
transnational companies in control of the drug market in Bhopal
were the chief beneficiaries of the disaster, even though there was
enough evidence to show that their drugs provided only temporary
relief. The herbal medicines and yoga used at the clinic are a direct
challenge to this power as the herbs are prepared locally, are cheap
and provide long-term benefits. All Western medicine that is used
is brought from the only non-profit collective in India. Patients
are able to choose their own care plan, treated as holistically as
possible, and their home lives as well as their health problems are
looked at. The clinic believes that Bhopal is not an isolated event and
that workers and communities are routinely poisoned all over the
world. They believe the only solution is the eventual elimination of
hazardous chemicals from the planet. They have found that there are
limits to 'modern' medicine in tackling industrial diseases, and that the
evolution of an appropriate system combining traditional and modern
medicine is necessary.

on house calls and to satellite and mobile clinics and provide health education. They have seen over 7000 patients in their first three months. They provided free health care in a period of crisis and continue to run the clinic with the aim of creating a permanent community controlled, primary care health clinic.

Networks

Personal and political networks are important. Having strong friendships and knowing we have personal support can play a big part in our health. In industrialised, capitalist societies, strong family and community networks have been lost. Many studies have shown that death rates or illnesses are two to four times higher among those who are socially isolated. One study found that after suffering a heart attack people with good social support are three times as likely to survive as those without (Wilkinson 2001). By building political networks without hierarchy we start to reduce the isolation of small groups and feel the political strength of many. It is also a good way of communicating, sharing skills and resources and thereby increasing our effectiveness. Activist Trauma is a recently created network in the UK. It is primarily for political activists who may have been injured during actions or other political activities and/or are struggling with mental health issues related to activism. They offer personal contacts as well as useful information around trauma and believe that supporting people who have been traumatised should be a central part of activism.

DIY health

Basic health care does not need a health professional. Knowledge of basic first aid to deal with a minor injury or knowledge of local herbs to relieve symptoms of a cold or other common ailments means people can look after most of their health issues. For centuries this information was passed down through communities and families, but has slowly been replaced by a dependence on medical professionals. We need to reclaim and share this knowledge through courses, reading or setting up groups. Education about health plays a crucial role here. It can be done within self-help groups, by an individual doing their own research or in the context of workshops put on in places like social centres or feminist health clinics.

Direct action and self-organisation

Today's society leaves most of us with little direct influence on many of the factors impacting on our health such as free trade agreements, pollution or war. Taking responsibility for our health includes finding new ways to challenges this situation and act collectively to influence the world around us. There are many examples of people self-organising to improve their lives and health, including food co-operatives to increase access to affordable healthy food, taking direct action against pollution or climate change, community allotments to grow fruit and vegetables and strengthen community support, and many more which are discussed in Chapter 17.

the way forward

In an ideal world there would be good quality housing, education that taught you to question and search for your own answers, land to grow food on, no repression and a supportive network of people. Some of this seems pretty unlikely but it doesn't mean that within the constraints of this society we cannot act to improve our health. Finding others is the first step and the next chapter offers many ideas of what groups can do to improve their health, be it starting a community allotment, a campaign against pollution or a health collective.

We would also like to see health care that improved health by working with people to help them meet their own needs, as defined by them. It is necessary to set up health centres that belong to the community rather than just health professionals. These centres need: resources that are made available to educate us on particular illnesses or ways of improving health; various types of health professionals working collectively and non-hierarchically; workshops on health issues; and a number of health collectives formed around different interests or needs. They could also have community cafes with cheap healthy food and healthy eating and cookery classes, community allotments, bicycle co-operatives and workshops, and walking or cycling groups. People can take action together against wider issues impacting on their lives. This is our vision of autonomous health which can make society healthier, not sicker.

6 how to manage our own health

Tash Gordon and Becs Griffiths

As the previous chapter showed, every part of our life has an impact on our health. Taking some personal responsibility for our health is an important part of taking control over our lives. It is very difficult to write a tick list for a way to achieve health which is more autonomous and self-managed, but what follows are some ideas on how to become less dependent on medical institutions that have fairly narrow ways of defining and treating disease. We can take more responsibility over our own health by eating more healthily, quitting smoking or drinking less alcohol. But we also need change at a wider social level – it's not just about taking individual responsibility. The emphasis on collective action means changing your own life, but also for and with others including those who need it the most.

There are a number of ways in which we can manage our own health and what is realistic depends on our personal situations and resources. Below we look at some examples including self-help groups, setting up health collectives, using natural remedies and dealing with medical institutions to your best advantage. What we explore is not intended to be a blueprint for autonomous health. We are not suggesting that these examples are all you need for better health. They are meant to show some steps towards ways of educating ourselves and supporting each other in becoming more informed and empowered – so we can decide for ourselves what works and what does not. To build the infrastructure and gain the skills needed for managing our own health is a long process and one which many of us, used to free and relatively easy to access health services, may find difficult to imagine. However, we will not be able to rely on public funded health systems for ever, and of course many already can't – those who cannot afford to pay, who have been refused asylum or those without papers, for example, are denied basic health services.

health collectives

Health collectives evolved from the women's health movement in the 1970s which was part of the broader women's liberation movement. This health movement emerged in the USA from the abortion reform activism of the late 1960s and became an important component of second wave feminism. A group of women in Los Angeles began to meet and do self-examination together (a technique that uses a plastic speculum to look at the cervix, the neck of the uterus, which feminists discovered was one of the most useful tools they had). Through self-examination women started to be more than passive consumers of health care as they learned more about their own personal cycles and bodies. This learning and teaching went beyond individualistic models of health care as it was done in small groups, emphasising the need for collective support and directly challenging the physician's role as expert. They taught breast self-examination, basic physiology and self-care, and provided critiques of medical studies, so women were able to make informed choices about their diagnosis and treatment – from having common infections like thrush to more serious illnesses like cancer.

These ideas spread around the USA and self-help clinics were set up enabling thousands of women to access them. The clinics in the USA also became abortion clinics after abortion was legalised in 1972. Eventually, some of these clinics formed

Box 6.1 Association of Radical Midwives

In the UK in 1976 two student midwives, who were frustrated and disappointed with the increasing medicalisation and intervention in maternity care, began meeting regularly for mutual support. They were soon joined by others and eventually set up the Association of Radical Midwives that still exists today. They are primarily a support group for people having difficulty getting or giving good, personalised midwifery care, which they think is predominantly lacking in the downsized NHS. They started the journal Midwifery Matters, which is now a respected journal mainly written by members themselves, and is on the shelves of a large number of midwifery schools and medical libraries.

Box 6.2   A personal reflection from a member of a Brighton feminist health collective

I have been part of a feminist health collective for five years in Brighton although the collective has taken different forms over the years. It has been a crucial transformative experience both politically and personally. The group evolved from a larger group of feminists who were meeting and were involved in broader anti-capitalist politics. Someone from the group gave a workshop on the history and politics of feminist health and the politics of menstruation and it all began from there. Many of us were amazed that we knew so little about our own bodies and how distorted the information about our own bodies was that we had been given throughout our lives. The collective started with 15 women and we met weekly in each other's houses, eating food together. Each session ranged from a couple of hours to a whole night and this continued for about a year. We looked at: the history of why so little information is given to women about their bodies and what kind of information this is, our experiences with the medical institutions, charting menstrual cycles, self-exam, ideas around our sexuality, breast health, abortion, pregnancy, AIDS, STDs. We had an allotment for several years on which we started to learn to plant, grow and harvest medical herbs together. One of the most important things I learned is to never solely rely on information given to me from mainstream medical texts or doctors. This is far from objective and contains certain motivations behind why and what information is presented. This was easier to grasp when everyone in the collective did their own research about a particular subject and, once again, we would be surprised at how we had accepted certain ideas and not challenged them.

We started to have monthly open workshops at a local social centre. They were about five hours long and we advertised them in blocks of two or three, explaining what each workshop would be about. From these sessions there evolved another health collective and this still continues. I think that this was important as people who were committed were able to go into topics in more depth as continuity was guaranteed. It also meant that there was a bigger

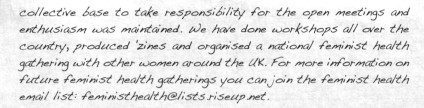

collective base to take responsibility for the open meetings and enthusiasm was maintained. We have done workshops all over the country, produced 'zines and organised a national feminist health gathering with other women around the UK. For more information on future feminist health gatherings you can join the feminist health email list: feministhealth@lists.riseup.net.

the 'Federation of Feminist Health Centres' network, which still comprises clinics scattered around the USA. In addition to setting up clinics women were active in other areas of women's health, such as sterilisation abuse, which happened particularly amongst poor and black women, lesbian health care, birthing practices and midwifery, access to alternative methods of contraception and workplace issues. As a result, many grassroots organisations were established to campaign and challenge the system. Some organisations became professionalised with paid jobs and an emphasis on single issues and in doing so lost their original radical critique. These ideas spread around Europe and health collectives sprang up in numerous places, although there was an emphasis on trying to reform the available free health care rather than creating distinct alternatives.

self-help groups

Setting up health collectives or self-help groups can create different networks of support as well as sharing skills and disseminating information. They can provide practical and emotional support in a mutually beneficial way and help people find collective solutions. Often, the greatest benefits are through active involvement in the organising of the self-help group. Below are some examples.

Action medics

Action medic groups have been created to meet the specific needs of the activist community that arise in a protest situation and also to provide that care in a way which is in line with the politics of autonomous, self-managed, protests. Current groups include the Black Cross Health Collective, Medical Activists of New York,

Boston Area Liberation Medic (BALM) squad and the Bay Area Street Medics in the USA. One was recently formed in the UK called Action Medics who provided medical support during the G8 summit in 2005.

These networks of activists have medical skills ranging from first aid to qualified doctors and provide ongoing training for new members. Most are volunteer run, non-hierarchical and consensus led with clear policies not to give information to the police, and with many of the medics having an activist background. Their main aim is to ensure that there are trained people during protests and demonstrations. They work as part of a team and are able to deal with emergency first aid situations. The collectives organise courses covering basic first aid and situations specific to protest, such as CS gas neutralisation. Some collectives have websites which feature basic first aid kit lists and basic first aid, including what to do when someone is in shock, has been sprayed with CS gas, has external and internal bleeding, has fractured or broken bones or has concussion. Other collectives have a wider role, like BALM, which runs free health clinics for progressive causes and carries out health care related political actions.

Another important recent strand of work has been the establishment of groups working to recognise, support and raise awareness about mental health issues related to activism. Providing safe spaces for recovery at actions and protest camps is one important part of this. The groups also work to understand how to recognise the warning signs of conditions such as post-traumatic stress disorder (PTSD). These groups do not set themselves up as experts and do not attempt to counsel people with serious illness. However they believe that establishing a support network is one way of challenging the stigma attached to these problems. Choosing a therapist sympathetic to your political cause can be key to recovery. Some people have found that rather than being helped to come to terms with the injuries inflicted on them by police brutality, they are asked to question what made them want to protest in the first place. The information collected on the activist-trauma website is useful for anyone who faces violence or repression wherever it comes from. It states: 'Supporting people who have been traumatised should be a central part of our lives, for without support and solidarity we can be easily picked off. This is not exciting or glamorous, it's hard work. However it can be rewarding, interesting and have very positive results. We may sometimes feel powerless in the face of all their power but we CAN help each other' (see www.activist-trauma.net).

Mad Pride

Mad Pride was set up as a grassroots network by people with negative experiences of the mental health system. Their aim is to campaign to end discrimination against psychiatric patients – an idea that came out of the 1997 Gay Pride festival in London. A few survivors of the mental health system thought they could put on a similar event and so got together and started to organise. They set up a non-profit company to develop Mad Pride, and in 1999 organised a series of gigs and concerts. The group produced an anthology *Mad Pride: A Celebration of Mad Culture* (2001) with personal stories of how people experience their 'madness'. The main objectives of Mad Pride were consciousness raising, educating the public about mental health issues, promoting more positive images of mental health, and countering discrimination and prejudice towards people who have experienced mental distress. They reclaimed the word 'mad' and spoke out about their traumatic experiences within the mental health system. They campaigned and held demonstrations against the proposed 1999 Mental Health Bill in the UK which included compulsory medical treatment in the community and increasing the time people could be held if sectioned.

Mad Pride is no longer active but different projects that were inspired by Mad Pride have sprung up, such as Mad Chicks, Mad Brighton, Chipamunka and Madnotbad. Mad Chicks focuses on issues specific to women mental health users, challenging discrimination and misinformation around mental health. They have put on one event with workshops and debates, and their website is full of information and links. Mad Brighton was a week of workshops, a spoken word night and art exhibition by a group of people who wanted to challenge the notion of madness and discuss ideas around so-called 'mental health'. Chipmunka is a patient driven publishing house of books written about madness and people's personal experiences. The Madnotbad website is a space for people affected by mental distress to share experiences of misdiagnosis, mistreatment and neglect within the mental health system.

How to set up a health collective/self-help group

This is a generic list that could be useful in setting up many kinds of self-help groups or collectives:

☆ You need to decide who you would like to be part of your collective or how you could get people to join. Advertise in appropriate place like cafes, social centres, community centres, postings on websites, newsletters or magazines.

☆ Do you want to have a close, small intimate group or bigger, more open meetings? A mixture of both can work. You may want to create spaces where you can advertise and invite lots of people but you need to be aware that by always getting new people continuity of conversation can be difficult. You need continuity to explore different issues and this really only works when you have a core group of people that share similar ideas and trust each other.

★ You can set up monthly drop-in sessions that you can advertise. You need to look for quiet places where you will not be disturbed, possibly social/community centres or private rooms in pubs. If you feel comfortable with everyone, then using each other's houses can be easier.

★ You need to decide what issues you want to cover. A starting point can sometimes be your own health experiences. Once the collective has been established you need to talk at length about what you each want from the group and realistically how committed people are. This might only happen after a few sessions once people get a feel about what the group is or could be like.

★ It's important to work out confidentiality. Basically what is said in the room stays there.

★ It can be really difficult to deal with a lot of different views as well as the fact that talking about health issues can be very emotive. Sessions can be a mixture of personal stories, sharing research and political debate.

★ There could be conflict and it is normally a good thing to talk about how you deal with conflict within the spaces you create. There will always be political debates and disagreements but it's good to remember to try and not make it too personal.

★ You can think of what you would like to do in the future: make pamphlets/ zines on your experiences or different topics, network with similar groups, put on workshops or other educational projects or whatever you may be creative enough to think of.

using natural remedies

Many of us have lost the ability to look after ourselves when our immune systems are down using simple remedies like ginger tea or fresh garlic. There is so much knowledge that we can reclaim starting with learning about the medicinal value of the common 'weeds' around us or the basic nutrition of fresh food. Wildcrafting is the practice of harvesting plants from their natural, or 'wild' habitats, for food, medicinal or other purposes. It doesn't have to be in the wild – there are many medicinal herbs that can be found growing out of pavement cracks and on urban wild patches, such as around allotments. You will be surprised by how many common herbs are growing all around us. Elderflowers, hawthorn berries, nettle, shepherd's purse, mullein, dandelion, burdock, borage, calendula, lavender, rosemary, to name a few herbs that grow everywhere in northern Europe. We can use these herbs and start to understand their benefits by learning about where they grow, how to pick them, dry them and then how to use them.

It's important to remember to never use plants growing within 10 feet of roadsides, along train tracks, near power lines and fences, or close to cultivated fields and fruit orchards (unless it is organic cultivation). The herbs will be affected by chemical contamination through vehicle emissions, creosote, herbicides and pesticides. Also take care when harvesting plants growing near or in water – there may be agricultural or industrial run-off upstream. You should choose the plants carefully ensuring that they are exactly what you think they are and are not endangered. Use a book and try and identify plants by their Latin names as you are less likely to make mistakes. Also, learn not to overharvest – one of the rules of thumb is the tree/plant should not look that different from when you started. Wild and cultivated 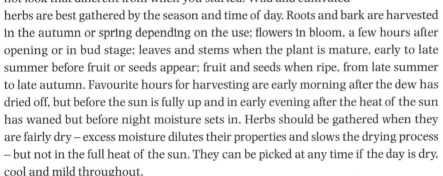 herbs are best gathered by the season and time of day. Roots and bark are harvested in the autumn or spring depending on the use; flowers in bloom, a few hours after opening or in bud stage; leaves and stems when the plant is mature, early to late summer before fruit or seeds appear; fruit and seeds when ripe, from late summer to late autumn. Favourite hours for harvesting are early morning after the dew has dried off, but before the sun is fully up and in early evening after the heat of the sun has waned but before night moisture sets in. Herbs should be gathered when they are fairly dry – excess moisture dilutes their properties and slows the drying process – but not in the full heat of the sun. They can be picked at any time if the day is dry, cool and mild throughout.

Exposure to air, heat, light and moisture are the main damaging factors for herbs, whether fresh, dried or in preparations. Fresh herbs should be used within 24 hours of harvesting, unless they are to be dried, in which case they should be prepared for drying within a few hours of harvesting. The first choice for storing dried herbs is an opaque glass or ceramic jar with a tight fitting lid. Metal tins, wood and cardboard boxes, lined with wax or craft paper, can also be used. Plastics are the least favourable containers for long time storage; the herbs cannot breathe and will often take on the taste of the plastic. They should be stored in a dry, cool space. You can store leaves, flowers and tender stems for up to one year, and roots, seeds, dried berries and bark for up to two to three years.

Herbal medicine can be complicated and the powerful effects of herbs should never be underestimated. Some plants can cause severe kidney and liver damage and possibly be fatal. It is possible to treat yourself for anything, but it's obviously better to start treating simpler things and then gradually build up your knowledge of each herb and its effects. Often the effects of the herbs can be very specific to the person. Despite this, there are some well known recipes that are often effective, normally relieving symptoms more than tackling the root cause of the condition. The simplest way to use the herbs is to make herbal teas. Take a pinch of one or more herbs, put them in hot water and leave for about 15 minutes for the active chemicals to infuse out.

What follows are a couple of simple recipes. For more complicated ones you will need to do the research and consult a medical herbalist. More holistic approaches to health emphasise a dialogue between the person being treated and the specialist. As with consultations with any expert, the more informed you are, the better you can use the information in the way that is right for you and your body.

Hayfever mix

Make a herbal infusion of chamomile (*Chamomila recutita*), elderflowers (*Sambucus nigra*), eyebright (*Euphrasia officinalis*) and nettle (*Urtica dioica*). You should start drinking this tea as early as possible, preferably before hayfever season starts. Take the tea two or three times a day, every day.

For eye soothers you can use eyebright tea in an eyebath. Infuse it like a herbal tea and let it cool. Drain the herb and soak cotton wool pads in the infusion and put them on your eyes for at least 15 minutes. It is particularly useful where there is a lot of discharge from the eyes.

Flu/cold

For a good 'cold tea' combine equal parts of elder (*Sambucus nigra*), peppermint (*Mentha piperita*), and yarrow (*Achillea millefolium*) and steep one to two teaspoons of the mixture in one cup of hot water. Take it hot just before going to bed. This will induce a sweat, and if the cold is caught early enough, may stop it altogether. Even if it is too late for this it will still be very useful. This tea can help the body handle fever and reduce aches, congestion and inflammation. It may be taken with a pinch of mixed spice and a little honey to soothe a painful throat.

You can make a warming tea throughout the day using cinnamon (*Cinnamomum zeylanicum*), cayenne (*Capsicum minimum*) and chopped fresh ginger (*Zingiber officinale*).

Use inhalations of chamomile, eucalyptus or thyme essential oils to help loosen mucus and heal the throat, nasal passages and bronchial tubes. Boil some water and put in a bowl with a few drops of each oil and then cover your head with a towel, lean over the bowl and inhale.

Garlic helps to detoxify the body. One chopped raw garlic clove on toast or in other meals can help strengthen the immune system.

Important! If you:

★ are pregnant, breast feeding or plan to become pregnant
★ have a long-standing illness
★ have undiagnosed health problems
★ are on medication

... seek advice from a medical herbalist first. If taken as directed, herbal and homoeopathic remedies are very safe, but they are powerful medicines. Do not use them lightly and do not exceed the recommended dose. When self-prescribing, be aware that you are responsible for your own actions and watch carefully how you react to the remedies. If you notice any adverse reaction, stop taking the remedy and find a herbalist to get more information.

dealing with medical institutions

Recognising the limits of self-managed health care is important, and in an emergency or with a serious health problem you may need to seek support beyond your networks. In the UK free medical care is supplied by the National Health Service which on the whole provides Western scientific diagnoses treated with allopathic (chemical) medicine. There are institutionalised hierarchies in this kind of medical system – those who have been ignored by a doctor or been made to feel stupid will know this. But there are also some inspiring and dedicated health professionals who aim to provide medical care which respects an individual's needs and desires. An awareness of how to best access the care you need is important. You have a right to play an active role in your care – the fact that you are dealing with a professional does not discount your opinion. Below are some pointers of how to get the most out of a health professional.

Getting the most out of a health professional

1. Find a health professional you trust and get on with. You need to have some belief and confidence in the type of treatments they offer and to be open with them. Either they should understand the way you choose to live your life or be non-judgemental about it. It is worth (services and funds permitting) finding the right person for you.

2. Be clear why you are going to see them. Do you have a new problem you are worried about? Has an old problem got worse? Do you suddenly feel unable to continue with long-standing symptoms? Do you have ideas about what it could be and do these worry you?

3. Think about what your expectations are. Are they realistic and fair? What do you know about the treatments or advice you do and don't want? The more informed you are, the more active you can be in planning your care.

4. Once you have established some trust, be open. The wider the understanding health professionals have of you, the better care they can provide. Be ready to try things you might not have thought about, as long as they can provide reasonable grounds for trying them.

5. Take the time you need. Maybe you need a few consultations before you establish trust or a good understanding of the problem. If you want to go away and think about things or talk them through with others, this should always be acceptable.

6. Make sure you understand. Clarification should always be sought if things feel unclear. Written information can be helpful to look through in your own time.

7. Ensure the treatment plan is a joint one. Unless you can see the point of the treatment and think that it is feasible in your current situation you may not be able to follow it through. The health professional is there to help you find the right treatment for you, not dictate it.

8. Have plans for future care. Know what to do if things get worse and if you need to be seen again or can manage your own care.

9. Take responsibility. If a physiotherapist has advised regular stretches – do them. If you know lots of late nights make you worse, try and avoid them. It's your health and you can go to as many appointments as you like but unless you are prepared to take care of yourself, no one else can.

10. Treat health professionals like a human beings. Yes, they are human and have good days and bad days and they appreciate a smile. It can be a demanding line of work, with a lot of people offloading problems onto them. Small things, such as asking how they are or saying thanks, make a big difference.

ways to better health...

This chapter, along with the previous one, hopes to inspire and give ideas about how we can organise collectively to have more control over our health – both by being more aware of how the society we live in interacts with our health and by giving examples of what we can do to challenge this. Our needs and capabilities will vary between individuals and over time so we need to be creative and open to new ways of tackling health issues and taking some control over them. A realistic attitude about the limitations of a society, where we face many barriers to our well-being, is important. Some problems will not be overcome by self-help groups or food co-operatives and allotments but only by significant social change. Inequality and oppression aren't easy to eradicate, but we can act to challenge and hopefully reduce them. In the same way, we can't simply do away with ill health, but we can do more to decrease our chance of it or speed our recovery.

Tash Gordon lives in Leeds, UK where she works in an inner-city GP practice. She does voluntary work with homeless and asylum seeker communities, runs an acupuncture clinic and is also involved in the Common Place – an autonomous social centre. Becs Griffiths has been part of a feminist health collective in Brighton, UK for the last five years to whom she is extremely grateful for all their wise advice and support. She is also involved in the Cowley Club, a social centre in Brighton. Becs was involved in the national feminist health gatherings in 2004 and 2007 and Tash was involved in 2007. The chapter draws upon many people's experiences and ideas. With particular thanks to Brighton Women's Health Collective.

resources

Books

Self-help, self-exploration

Boston Women's Health Book Collective (1989). *The New 'Our Bodies, Ourselves'*. New York: Simon & Schuster.

Chalker, Rebecca (2000). *The Clitoral Truth.* New York: Seven Stories Press.

Federation of Feminist Women's Health Centres (1981). *New View of Women's Body.* Feminist Health Press. New York: Simon & Schuster.

Shodini Collective (1997). *Touch Me, Touch Me Not: Women, Plants and Healing.* New Delhi: Kali for Women.

Weed, Susan (1996). *Breast Cancer? Breast Health! The Wise Woman Way.* Woodstock, NY: Ash Tree Publishing.

General health

Douglas, J. (1992). 'Black Women's Health Matters: Putting Black Women on the Research Agenda'. In H. Roberts (ed.) *Women's Health Matters*. London: Routledge. 45–60.

Ehrenreich, Barbara and Deidre English (1973). *Witches, Midwives and Healers*, London: Compendium.

Ehrenreich, B. and D. English (1974). *Complaints and Disorders: The Sexual Politics of Sickness*. New York and London: Feminist Press.

Ehrenreich, B. and D. English (1976). *Witches, Midwives and Nurses*. Old Westbury, NY: Feminist Press.

Hartmann, B. (1995). *The Global Politics of Population Control*. Boston, Mass.: South End Press.

Heath, I. (2005). 'Who Needs Health Care – The Well or the Sick'. *British Medical Journal* 330: 954–6.

Iheanacho, I. (2006). 'Drug Trials – the Dark Side: This World'. *British Medical Journal* 332: 1039.

Illich, I. (1976). *Limits to Medicine*. London: Marion Boyars.

Johanson, R., N. Newburn and A. Macfarlane (2002). 'Has the Medicalisation of Childbirth Gone Too Far?' *British Medical Journal* 324: 892–5.

Kawachi, I. and B. P. Kennedy (1997). 'The Relationship of Income Inequality to Mortality – Does the Choice of Indicator Matter?', *Social Science and Medicine* 45. Amsterdam: Elsevier Science Ltd.

Laws, S. (1991). *Issues of Blood. Politics of Menstruation*. Basingstoke: Macmillan.

Leslie, E., B. Watson, T. Curtis and R. Dellan (eds) (2001). *Mad Pride: A Celebration of Mad Culture*. London: Spare Change Books.

Money, M. (ed.) (1993). *Health and Community*. Totnes: Green Books Ltd.

Moynihan, R. and A. Cassells (2005). *Selling Sickness: How the World's Biggest Pharmaceutical Companies are Turning Us All into Patients*. New York: Nation Books.

Moynihan, R., I. Heath and D. Henry (2002). 'Selling Sickness: The Pharmaceutical Industry and Disease Mongering'. *British Medical Journal* 324: 886–91.

Nissim, R. (1986). *Natural Healing in Gynaecology*. New York: Pandora Press.

Thomas, Keith (1971). *Religion and the Decline of Magic. Studies in Popular Beliefs in Sixteenth and Seventeenth Century England*. Oxford: Oxford University Press.

UNDP (2006). *United Nations Development Programme Human Development Report* (http://hdr.undp.org/hdr2006).

Vincent, P. (2002). *Babycatcher*. New York: Scribner.

Wilkinson, R.G. (1992). 'Income Distribution and Life Expectancy'. *British Medical Journal* 24: 63–72.

Wilkinson, R.G. (2001). *Mind the Gap: Hierarchies, Health, and Human Evolution*. New Haven: Yale University Press.

The witch hunts

Frederici, Silvia (2004). *Caliban and the Witch: Women, the Body and Primitive Accumulation.* New York: Autonomedia.

Lady Stardust (2006). *The Witch Hunts in Europe 1530–1690,* http://escanda.org/RWG/ Texts/witchtrials.html

Websites

Campaigns and self-help collectives

Activist Trauma www.activist-trauma.net

Baby Milk Action www.babymilkaction.org

Bay Area Radical Health Collective www.barhc.w2c.net

Black Cross www.blackcrosscollective.org

Boston Area Liberation Medic Squad   www.bostoncoop.net/balm

Common Ground Health Collective www.cghc.org

Medical Activists of New York http://takethestreets.org

Sambhavna Trust, Bhopal www.bhopal.org

UK Action Medics www.actionmedics.org.uk/index.html

Mental health

Chipmunka www.chipmunkapublishing.com

Icarus Project www.theicarusproject.net

Mad Chicks www.mad-chicks.org.uk/index.htm

Mad Pride www.ctono.freeserve.co.uk

Madnotbad www.madnotbad.co.uk

Natural remedies

Herbal medicine www.susanweed.com

United Plant Savers www.unitedplantsavers.org

Women's health

Bloodsisters www.bloodsisters.org

Handmedown Distribution vivavocewimmin@yahoo.co.uk (Email for books on feminist health.)

Sister Zeus www.geocities.com/sister_zeus

7 why we still have a lot to learn

The Trapese Collective

> Tell me, and I forget. Show me, and I remember. Involve me, and I understand.
> (Chinese proverb)

Education, and in particular popular education, is vital to respond to the ecological, social and climatic crises we face and to achieve meaningful radical social change. An education where we relearn co-operation and responsibility that is critically reflective but creatively looks forward – an education that is popular, of and from the people. There are many examples of groups that organise their own worlds without experts and professionals, challenge their enemies and build movements for change. What we outline here is what is known as popular, liberatory or radical education which aims at getting people to understand the world around them, so they can take back control collectively, intervene in it, and transform it. This chapter looks at the importance of education in bringing about social change, and indeed how social movements for change have popular education at their core.

so what does popular education mean?

The word 'popular' can mean many things and has been mobilised by the right as well as the left. There is no single political project behind the methods of popular education. It has been used by all sorts of people including revolutionary guerrillas, feminists, and adult educators; all with different aims. Development practitioners from organisations such as the World Bank, for example, increasingly use popular or participatory education to co-opt, manipulate and influence communities to secure particular versions of development. Yet it is important to promote and reclaim some of the more radical strands of popular education which are rooted in defiance ('we are not going to take this anymore'), and struggle ('we want to change things'), and geared towards change ('how do we get out of this mess'), while promoting solidarity ('your struggle is our struggle').

The Popular Education Forum of Scotland (Crowther et al 1999) defines popular education as:

1. Rooted in the real interests and struggles of ordinary people.
2. Overtly political and critical of the status quo.
3. Committed to progressive social and political change.
4. A curriculum which comes out of the concrete experience and material interests of people in communities of resistance and struggle.
5. A pedagogy which is collective, primarily focused on group rather than individual learning and development.
6. Attempts to forge a direct link between education and social action.

One of the activities that we have used in our workshops to start thinking about how we learn is to look at positive and negative learning experiences. People have told us that negative experiences are characterised by fear, discipline, constant assessment, humiliation, being bullied or bored, and unenthusiastic teachers.

Positive experiences, on the other hand, are often those that are creative, interactive, student led, interesting, when learners are given responsibility, and take place in a supportive and friendly environment. Although many teachers in the state sector use participatory and progressive teaching methods, state funded education remains constrained by large class sizes, national curricula and targets. Table 7.1 (*overleaf*) outlines briefly some of the main differences between the overarching aims of formalised education in schools and universities, and popular education.

key aspects of popular education

1. A commitment to transformation and solidarity

Popular educators are not experts who sit on the sidelines; they participate in social movements, be they literacy campaigns, teach-ins about globalisation or by participating in actions. Solidarity means being on the side of the marginalised and working with agendas and goals chosen by those affected, not outside agencies. Charity and aid can provide temporary relief but are rarely designed to break the chains of dependency and encourage people to run their own affairs. At the heart of popular education is a desire not just to understand the world, but to empower people so they can change it. By identifying and exposing power relations, we can

Table 7.1   Comparing formal and popular education

Formal-state	*Popular-participatory-liberatory*
Why? To gain basic skills and teach acceptance of authority and preparation for participation in waged based work and consumerism.	**Why?** To raise critical consciousness, link with campaigns and action, and promote social justice and solidarity.
How? Learners receive knowledge from teachers, there is an emphasis on the end result, qualifications, exams and competitive grading systems.	**How?** Participants are active in how and what they learn. Hierarchies are challenged. Educators understand learning occurs in many different ways and employ a variety of techniques to build collective knowledge.
What is taught? Rational, fact based, information, learning skills for business and efficiency.	**What is taught?** Exploration of alternatives and radical solutions. Values emotional responses.

Box 7.1   Mujeres Libres (Free Women of Spain)

In the late 1930s in Spain, Mujeres Libres (Free Women) mobilised over 20,000 women and developed an extensive network of activities to empower individuals and build community. The movement saw education as central to releasing women's potential and to free them from the 'triple enslavement of ignorance, enslavement as woman and enslavement as a worker'. Classes were organised through cities and neighbourhoods. In 1938 in Barcelona alone, between 600 and 800 women were attending classes daily to *capacitar* (empower or prepare) them for a 'more just social order'. They organised autonomously from men, arguing that only through self-directed action would women come to see themselves as competent, capable and able to participate in the revolutionary movement. Classes ranged from basic literacy to social history, law, technical skills and languages. They spread their message further through books and pamphlets and speaking tours. Virtually all the activists were self-taught, putting into practice theories of direct action and 'learning by doing'.

Source: Adapted from Acklesberg 1991, *Free Women of Spain.*

begin to develop an understanding of how we might challenge ways of organising social and economic life which perpetuate injustice, at whatever scale.

This type of education is not just about explaining external problems, but also confronting each other and the social roles we have adopted. To make 'other possible worlds', we must also change ourselves and learn to listen to the experiences of others. It's not enough to understand how we think power works 'out there' if we overlook our role in reproducing power. Asking questions rather than providing answers is fundamental: What do we accept and reject? How do we pass on systems of domination be they class, gender, ethnicity or sexuality? How can we challenge these within ourselves? This way of teaching and learning is challenging and requires effort on both sides. Paulo Freire called it the 'practice of freedom' and talked of the dialectical relationship between the oppressor and the oppressed. This is what Freire meant by *conscientizacion*, through which we recognise our presence in the world, and, rather than adapting or adhering to social norms, we realise our potential to intervene and challenge inequality.

2. Learning our own histories not his-story

Although there is always at least two sides to every story, the vast majority of official history is exactly that – 'his-story', written by the literate educated few, mainly men, not by peasants, workers or women. We are taught about leaders of world wars and histories of great scientists, but not much about the silent millions who struggle daily for justice. These are the ordinary people doing extraordinary things who are the invisible makers of history. When they do appear, they are often portrayed as violent extremists. Few learn about the Haymarket martyrs in nineteenth-century Chicago who fought for an eight-hour day, the nineteenth-century Luddites who challenged the factory system during the Industrial Revolution, or the women who recently occupied the Shell platforms in Ogoniland, Nigeria. Many of these stories are not told because people could not read or write, or did not have any means to record events and communicate with a wider audience. They are not recorded by historians because they evoke the dangerous idea that ordinary people can act collectively and do it themselves. Talking about a proposed gas pipeline in County Mayo, Ireland, a campaigner reflected:

> A generation ago we could not have resisted this pipeline, because we could not read and write – we wouldn't have been able to respond to what Shell were saying and doing or fight them in the courts. Now we can fight Shell on the same level and they don't know what to do. (Vincent McGrath, Shell to Sea Campaign, interview with authors, June 2005)

It is important to relearn our own hidden histories of struggle, they exist everywhere and can be uncovered. They can help to dispel apathy ('it's not worth it') and powerlessness ('it's too overwhelming'). Learning these lessons shows us that most of our freedoms and improvements, which we value in our lives today, have been fought for and won through collective and sustained action by people like ourselves, not great leaders. Oral history projects that engage with members of a community and record their memories, and walks that visit sites of historical interest, of uprisings and old ways of life are two ways of relearning and connecting these forgotten histories.

3. Starting from daily reality

Any project should begin by looking for connections between problems and people's everyday lives, not a preconceived idea of this reality. Popular education is about avoiding judging people and encouraging people to express themselves, in their own way. It is not about learning lists of facts, but looking at where people find themselves and how they understand what's going on around them.

Many believe that learning happens best where there is affinity between the educator and the participants – and when common experiences can be used as the material to be studied. In his work on popular education in the El Salvadorian revolution, John Hammond observed that the teachers in the National Liberation Front (an army largely made up of illiterate peasantry) were generally combatants who had only recently learnt to read or write themselves. The biggest challenge can be building bridges between disparate worlds. Talking about television or football and finding things in common can be ways to start a conversation. It takes time to connect with people, and respect and trust are the keys to positive learning.

4. Learning together as equals

Popular education methods are designed to increase participation and break down the hierarchy between educator/teacher and participant/learner. The educators and those they are working with collectively own the process, ideally deciding the curriculum and determining the outcome of the action to be taken. Whilst in many contexts educators are seen as experts who can provide quick fixes, popular education has an explicit aim to reduce dependency between educators and those they engage with. Radical educator, Myles Horton, would tell his students at Highlander that if he gave them an easy answer today, what would stop them coming back tomorrow and asking him again? He argues that groups trying to find a way out of a problem are often the most capable of experimenting with possible solutions and should be encouraged to do so.

Box 7.2 Highlander Folk School, USA. 'The people always know best'

The Highlander School emerged from the needs of various social movements in 1930s Tennessee. It initially got involved with the labour movement, helping workers to organise. By the mid 1960s it was central to the civil rights movement organising literacy classes in poor black communities, teaching them to read and therefore enabling them to register to vote. They started classes by reading the United Nations Universal Declaration of Human Rights, the powerful language of 'all men being equal', encouraging those who attended citizenship schools to demand more than just the right to vote. Participants included Martin Luther King and Rosa Parks, the latter sparking a desegregation movement by being the first black woman to refuse to give up her seat on a bus. Myles Horton, one of the founders, has become famous for his approach which argues that ordinary people have the ability to understand and positively change their own lives. The school continues today; its mission is 'to build strong and successful social-change activism and community organizing led by the people who suffer most from the injustices of society'.

5. Getting out of the classroom

A critique of the powers and rules we live by cannot flourish when learning only happens within the official institutions and places controlled and funded by those in power. The state control of schools and compulsory education is not inevitable, nor does it reflect a widely articulated need. However, it has become all encompassing. The school forms the ideology, patriotism and social structure of the modern nation state. Free, compulsory education is now based on the assumptions that the state has the responsibility to educate all its citizens, the right to force parents to send their children to school, to impose taxes on the entire community to school their children and to determine the nature of the education on offer.

One particular issue is the creeping influence of corporations on our education – through private academies, but also through sponsorship of learning materials, research, and even food and entertainment. Whilst we enjoy a 'free education' (i.e.

we generally don't have to pay) the influence of private corporations in the delivery of curricula as well as in schools' facilities is increasingly a cause for concern. Teachers are ever more limited in what they can teach by the national curriculum, and there is more compulsory testing from a younger age. Students are taught conformity to values chosen by government and increasingly big business. A recent report highlighted the links between universities and the oil industry:

> Through its sponsorship of new buildings, equipment, professorships and research posts, the oil and gas industry has 'captured' the allegiance of some of Britain's leading universities. As a result, universities are helping to lock us in to a fossil fuel future. (Muttitt 2003, 2)

Getting out of the classroom and institutionalised learning environments is a key part of rethinking learning about everyday life – outside encounters, street life, listening to somebody, at home, within the community are all places of learning that gives us valuable social skills and rounds our knowledge. This type of learning is also about challenging education's negative associations and making learning passionate, interesting and challenging. People learn everywhere and using social and cultural events, music, food and film is a good way to reach out to people who may not come to a talk or workshop. Experiments in education beyond formal schools include 'Schools Without Walls', such as the Parkway Program in Philadelphia where the whole city is used as a resource.

6. Inspiring social change

Discussing important subjects, such as climate change, can be depressing and can leave us with feelings of despair and doom. Rather than avoid talking about them we can look at ways to deal with this. Firstly, we can identify a number of common barriers to changing attitudes or behaviour:

★ Apathy, 'I can't be bothered' or 'it doesn't affect me'.
★ Denial that the issue exists.
★ Feeling of powerlessness to do anything about the situation.
★ Feeling overwhelmed by the size and scale of the problem/issue.
★ Socio-economic time pressures and lack of support.

The way that learning happens can turn these attitudes around and help us turn our outrage and passion into practical steps for action, our dreams into realities. We can explore examples from the past where people have struggled and won and focus on workable alternatives. Practical tips for planning a workshop can help, such as identifying small achievable aims, breaking down issues into manageable chunks (see the next chapter for specific exercises), providing further resources, helping with action planning and campaign building.

Radical educators take on the responsibility to guide groups beyond common fears to reveal answers and possible escape routes to problems, laying possible options on the table. The art lies in the ability to make connections and establish bridges between people's everyday realities and what they start to think is possible in the future. Hence, inspirations for social change are presented slowly and gradually, with honest reflection, compromise and setbacks along the way.

Part of this learning experience is about sharing what is feasible, both in the here and now and other times and places. There are many workable ways of living that directly challenge the money economy, wage labour and ecological crises – many of which are discussed in this book – working co-operatives, community gardens, low impact living, direct action, autonomous spaces, independent media. On their own they may not seem much and are spread far and wide. But if they are gathered up and presented collectively they can provide excitement and hope and form a basis for a more creative, autonomous life.

popular education in action

There is a rich history that criss-crosses the world as people have struggled for freedom and against oppression. Popular education has flourished at times of big social upheavals, when people question the way the world is, and see a need to change their lives.

Educating the workers for freedom

The Industrial Revolution meant massive changes and new realities such as overcrowding, long working days and urban poverty. Working-class people in the UK did not have the right to formal education; in fact many educators and members of the aristocracy argued that education would confuse and agitate working people. Various associations were established to campaign against this injustice. Some authorities conceded that education for working people might be useful so long as it was devoted only to basic skills development. Associations struggling against these views developed their own forms of education – 'rag' magazines, study groups and community activities. Socialists of various affiliations struggled to educate themselves and those around them to understand and tackle the horrific new realities of life, whilst openly trying to develop class consciousness.

The book *The Ragged Trousered Philanthropist* (1918) by Robert Tressell is one famous example. It depicts the efforts of Owen, a firebrand socialist painter, trying to educate his reactionary pals about the evils of capitalism. A rich tradition exists ranging from the Labour colleges, the Correspondence societies during the revolutions in France and the USA, to later experiments such as co-operative colleges, Workers' Educational Associations, and adult education colleges such as Ruskin College in Oxford. Many of these presented a blueprint for transformation to a socialist society, based more or less on a Marxist-Leninist perspective. Alongside the workshop and the trade union, Marxist schools or workers' universities were set up. These sprang up across the world into the twentieth century, offering classes to workers in the basics of socialist thinking whilst also training professional international socialist activists and agitators, and becoming a focus for anti-communist surveillance and repression. Radical organising in working-class communities has continued through tenants' and claimants' unions, and in the UK through anti-poll tax unions drawing on these powerful roots.

Free schools

Many educational alternatives have been tried over the years, experimenting with radical education through free or progressive schools. Many had revolutionary potential, not just undermining state power, but also challenging ways of life and were seen as a real threat. For example, Spanish anarchist Francisco Ferrer was executed for plotting a military insurgency when he opened a school that was free from religious dogma. The high point for free schools was the New Schools movement in Europe in the mid twentieth century. Schools were based on voluntary attendance and children and teachers governed the school together. There was no compulsory curriculum, no streaming, no exams or head teacher. Instead libertarian ideas were promoted and there was a focus on creative learning and interaction between different ages and the outside world. Free schools exist throughout the world, such as Mirambika in India and Sudbury Valley School in the USA. British examples include Abbotsholme School in Staffordshire and Summerhill in Suffolk. While education is compulsory, schooling isn't; networks such as Education Otherwise and the Home Education Network provide support for the thousands of parents in the UK who choose to educate their children at home.

Struggles for independence

Popular education movements have played central roles in the struggles for independence in many colonised countries. In the twentieth century, socialist inspired nationalist struggles across Latin America and Africa used popular education to engage with the masses, challenge oppression, apartheid and colonialism. Liberatory educators in countries including Nicaragua, Granada, Cuba, El Salvador and South Africa set up educational programmes to mobilise the masses, especially the rural poor. In these revolutionary contexts popular schools flourished. 'People's Education for People's Power' in South Africa, for example, was a movement born in the mid 1980s in reaction to apartheid and was an explicit political and educational strategy to mobilise against the exploitation of the black population. It organised Street Law and Street Justice programmes and literacy and health workshops – these programmes were also subject to repression.

Latin America

One of the best known examples of popular education being used to challenge oppression and improving the lives of illiterate people is the work of Paulo Freire in Brazil. Working with landless peasants, he developed an innovative approach to literacy education believing it should mean much more than simply learning how to read and write. Freire argued that educators should also help people to analyse their situation. His students learned to read and write through discussion of basic problems they were experiencing themselves, such as no access to agricultural land. As the causes of their problems were considered, the students analysed and discussed what action could be taken to change their situation.

Radical popular education has recently seen a resurgence in Latin America, as people try to make sense of the current crisis brought about by 30 years of neoliberal economic policies. In Argentina after the 2001 economic crisis, Rondas de Pensamiento Autonomo (Roundtables for Autonomous Discussion) and open platforms in neighbourhood assemblies have become common features where people talk about the crisis and possible solutions. The Madres de Plaza de Mayo (The 'Mothers' who tirelessly campaign against the disappearance of many innocent people during Argentina's dirty war) have set up the Universidad Popular Madres de Plaza de Mayo on the Plaza del Congreso in the centre of Buenos Aires. This people's university, dedicated to popular education, houses Buenos Aires' best political bookshop, the literary cafe Osvaldo Bayer, a gallery and workshop space which holds classes, seminars and debates on topics from across Latin America. Since the establishment of the Venezuelan Bolivarian Republic in 2001 under Hugo Chavez, Bolivarian circles and local assemblies have spread to engage people in implementing decision making and the new constitution.

Education for global justice

Over the last ten years, the anti/alter-globalisation struggle has been a hotbed for popular education activity. A global summit of world leaders rarely passes without several teach-ins, counter-conferences and skill sharing events where activists and campaigners come together to inspire and inform each other about what they are attempting to understand and challenge. Groups and networks have emerged dedicated to producing and disseminating a huge amount of information on topics crucial to understanding our contemporary world: sweat shop labour, fair trade, immigration, war and militarisation, the effects of genetically modified organisms, neocolonialism and climate change. Hallmarks of such workshops are teaching horizontally and encouraging equal participation. Whilst big campaigns and mobilisations are often times for such educational outreach there are many social centres that provide space for ongoing autonomous education. La Prospe in Madrid has, since the mid 1970s, hosted Grupos D'Apprendizaje Collectiva (Groups for Collective Learning) on topics such as gender, globalisation, basic skills and literacy. As well as converging at global summits there are many international gatherings and seminars that all provide means of exchanging, building and networking ideas and experiences of different groups.

where now for popular education?

Social change will not be achieved by a small group of experts but will involve bringing people together on an equal basis. One of the issues that has faced the alter-globalisation movement is its need to communicate with wider audiences to get off the activist beaten track. Although popular education on its own is not enough, it is one way for people to engage themselves and their communities in these discussions, to begin to think of their needs and the possibilities that can be created.

In post-9/11 USA, Katz-Fishman and Scott argue that a climate of fear, hysteria and pseudo-patriotism has been created to control and contain dissent. They argue that:

> To prevent the fragmentation and break down of the community means organizing ongoing educational development among grassroots-low-income and student-scholar activist communities of all racial-ethnic-nationality groups and bringing people together on the basis of equality. (Katz-Fishman and Scott 2003)

This, they argue, must be done 'community by community and workplace by workplace'. We echo and support this call to action, to extend groups and networks of popular and radical education.

There is a widespread sense that something is not working. The illusion of infinite upward economic growth can't be maintained for much longer as natural resources become scarcer and the capacity for the planet to absorb waste becomes exhausted. There is therefore a potential for radical critiques to be articulated and developed. Popular education tools can help us do this in ways that make sense to people and to reveal why alternatives are possible and necessary. One great potential of popular education is that its participatory methods mean that 'activists' learn to make their ideas relevant and accessible. In a world where we all impact upon the lives of others, the boundary between who is the oppressor and the oppressed becomes increasingly confused. In the developed world, as consumers of the world's resources driving a system of global exploitation, we must teach ourselves about the impacts we have on the world, the role our governments play, how to take responsibility, and, most importantly, how we can take action to change this. An education that seeks to address unequal power relations and empower collective action is vital. The work of people over the centuries, with limited resources but with a passion for change, should be our inspiration.

8 how to inspire change through learning

The Trapese Collective

How things are taught is as important as what is taught in inspiring people to take action in their own lives. In these days of compassion overload we can't assume that any shocking statistic or distressing story will have any impact. Instead we need creative ways to think and learn about the problems we face.

In the year 2004–05 shortly after we formed our popular education collective Trapese, we carried out over 100 workshops, talks and quiz shows round the UK and Ireland exploring issues of the G8, climate change, debt and resistance. Since then we have continued our work exploring popular education methods as a way to support a range of campaigns. This chapter brings together practical advice which is based on our own experience, and from other groups who also use popular education as a tool for change.

Many of the activities and games mentioned in this chapter have been adapted from tried and tested methods of others doing similar work who made their resources available. What links the activities together is that they aim to create a collective understanding of problems, root causes and encourage people to take action, tapping into a desire for change. This is just a starter, there are many websites and books that expand all these ideas (see the resources section at the end of this chapter), but we believe that there is no better way to learn than by doing.

getting organised

Here are some of the stages in organising an event.

(a) Knowing the subject
   ☆ Choosing the theme should be the easy part, remember to have a clear focus for the workshop. Local issues can quickly be scaled up and connected with bigger questions.

⋆ Find out as much as possible about the participants – how often they meet, what their interests are, what level of awareness they may have about the topics you want to talk about.

⋆ While you don't need to be an expert, it's important to have some concrete facts as they will help give you credibility and confidence. Use books, films and websites, newspaper clippings and quotes from the radio, TV and films.

⋆ Research any existing campaigns and try to understand the arguments of all sides.

(b) Designing the workshop

⋆ Running a workshop with more than one person can really help practically – it also gives more variety.

⋆ Bear in mind that people normally retain more if they have an opportunity to discuss, question and digest. Less is more.

⋆ Remember that there are neither correct answers nor easy conclusions. The aim of the activities is to plant the seeds of questioning and encourage people to find out more for themselves.

⋆ Use a variety of different types of information – films, games, debates and allow free time for questions and informal discussion.

⋆ Include plans for action and possible future steps early on. Things often take longer than you imagine and it's depressing to hear all about a problem and then be left with no time to discuss what to do about it.

⋆ Allow time for breaks – in our experience, any more than one and a half hours and people will start to switch off.

(c) The practicalities

⋆ Getting people along can be the main challenge, look out for existing groups, unions, community groups/centres and spaces which have similar events.

⋆ Advertise as early and as widely as possible using posters, websites, email lists, etc. but also think about personal invitations, which can be most effective.

⋆ Set up all the equipment you need well in advance to avoid last-minute stress.

⋆ If space permits, arrange the chairs in a circle as people can see each other and there is no one at the front lecturing.

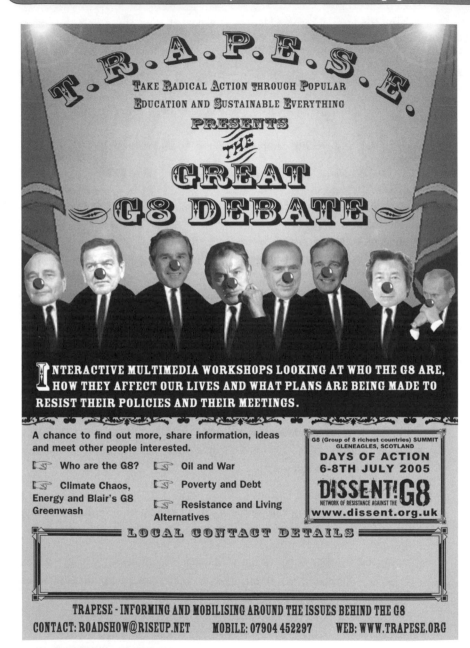

Figure 8.1   Poster advertising our workshops

Source: Andrew X.

    ★ Think of a method for people to give feedback and to exchange contact details.

    ★ Provide snacks and drinks.

    ★ Offer people the possibility of further sources of information, through either handouts or websites.

(d) Facilitating

    ★ Keep to an agreed time frame and explain the aims and structure of the workshop.

    ★ If you are friendly and respectful, then other people are more likely to follow your example.

    ★ Make a brief group agreement at the start – this can include things like everyone will turn off mobile phones, agree to listen to other people speaking, wait their turn, etc.

    ★ Ask people who haven't spoken if they would like to contribute.

    ★ Don't be afraid to admit that you don't know the answer. You can offer to find out or suggest that you find the answer together.

    ★ People learn best when they come to their own conclusions. The facilitator's role is to lead people through information, rather than presenting completed solutions. Ask questions and encourage participants to ask questions. For example, 'The way it works is...' can be replaced by 'Why do you think it works that way?' This may take a bit longer but it is more likely to be absorbed.

    ★ Use bright, colourful props and a range of media to draw people's attention. Dress appropriately to the group.

exercises for social change

Think of a really boring teacher at school. What made them boring? Were they monotonous, arrogant, bossy or stern? Think of some piece of information that really impacted on you. Why do you remember it? What struck you about it? How was it presented? Thinking about being a participant yourself will help you to plan a workshop. At the same time, remember people learn in different ways – through listening, writing, drawing, speaking and acting – so try and use a variety of senses. Many of these activities can be easily adapted to work on other topics.

1. Warm ups

Warm ups set the context of workshops and allow everyone to get to know each other. They can be more animated or calming depending on what you feel is appropriate for the group. Games can be a good way to create a participatory environment where everyone feels they can contribute. Some physical contact (being aware of different abilities and cultural sensitivities) can be a good way to relax people and break down personal boundaries. Go round the room and ask people to say their names, and if there is enough time ask them to add what they hope to get out of the workshop. This can help the facilitator pitch things accordingly.

Play a game

Before you do anything, try playing a short physical game – we all know many from our childhood, such as musical chairs, keeping a ball or balloon up in the air, or stuck in the mud.

Finding Common Ground

Aim: An icebreaker and a way to see how many similarities exist in the group's opinions.

Method: Everyone stands in a circle. Explain that when a statement is read out, if they AGREE they should take a small step forward. If they DON'T AGREE, stay put. No steps to be taken backwards. Statements should try and reflect the interests of the group and controversial or topical issues. for example:

- ☆ 'I think corporations are taking over our political processes'.
- ☆ 'This makes me angry'.
- ☆ 'I drink fair trade coffee/tea'.
- ☆ 'I don't think that's enough'.
- ☆ 'If more world leaders were women, the world would be a better place' etc.

Depending on the size of the group, with ten or so statements everyone should be in the centre of the room. At this stage you can all sit back down again. Alternatively, ask each member of the group to close their eyes and put out their hands into the middle of the (now very small) circle. Ask each person to take two other hands. When they open their eyes the task is to untangle the knot of hands.

Outcomes and tips: The tangle is also good for working together out of an apparently impossible mess. Be prepared to abandon it if it takes ages!

2. Collective learning

Before beginning an explanation about collective learning, ask people what they already know about it. One way of doing this is to ideastorm around an issue. Ask people to shout out what they know about something and write it up visibly so that everyone can see the ideas. Don't correct people at the time but make a mental note to come back to the point later.

Acronym game/Articulate

Aim: To jargon bust, build group understanding of terms and to gauge the existing knowledge of a group. The game introduces lots of background information and gets people working in teams.

Method: Write out some relevant acronyms or words on small cards. Divide the group into teams and divide the cards so they have roughly one per person. Ask groups to discuss the cards and work out what they are/mean/ do. Help if necessary. Each group then presents the acronym to the other groups without saying any of the words in the name. For example, if you have WTO you can't say the words 'World', 'Trade' or 'Organisation' in your description but something like, 'It's a global institution that makes rules about and removes barriers to trade'. Or an emotional response, 'It's the most damaging institution in the world and should be abolished'. The team which correctly guesses what the acronym stands for receives the card. Ask the group if they can explain the idea in more detail.

Some acronyms we have used include:

☆ PFIs (Private Finance Initiatives). Corporations investing in public services, such as hospitals and schools.
☆ IMF (International Monetary Fund). Lends money to developing countries; generally comes with conditions on market based reforms.
☆ WB (World Bank). Lends money to projects in developing countries, mainly focusing on large infrastructural projects like dams and roads.
☆ SAPs (Structural Adjustment Programmes). Conditions for IMF loans which involve liberalising the economy, deregulating and privatising industries.

Outcomes and tips: Demystifying complicated acronyms and terms is important to developing a critical awareness about our world. This game can last a long time so be prepared to cut it short in order to stick to your workshop plan. Any words, names or ideas can be used for this game and it is a good lead in to the Spidergram game (see below).

Spidergram (Mapping Climate Change)

Aim: To explore a topic visually, make connections between ideas and unpack cause and effect.

Method: In small groups, draw a box in the middle of a big piece of paper and write the big theme which you want to explore, e.g. 'climate change'. Ask people to think of things which directly cause this like 'flights', 'cars' and connect these to the centre with a line. Then think of problems or issues relating to these issues like 'pollution', 'asthma', 'traffic jams', etc. If linking to the acronym game, mentioned above, choose a couple of cards and ask people to draw connectors to other cards and arrange them.

Outcomes and tips: You'll soon build up a picture of connections like a spider's web. Ask the group which words have the most links. Make sure you go round and help groups.

What is happening?   *Why is it happening?*

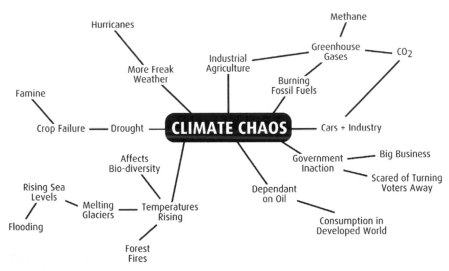

Figure 8.2   Spidergram

Source: Trapese Collective.

3. Using visual activities

It is often more striking to see something simply but visually than to listen to a long list of statistics. Physical and visual activities change the pace and dynamic of a workshop, which helps participants retain concentration. Look for possibilities of involving practical tasks, training or experiments in the workshop.

Chair game

Aim: A simple way to show the imbalance between the G8 countries and the rest of the world. This game can be modified to represent other imbalances or statistical information, e.g. debt, trade or carbon emissions.

Method: Ask ten volunteers to form a line with their chairs and to sit on them. You are going to ask a series of questions and in each question, one chair equates to 10 per cent of the total. During the game, people will move along the chairs according to their allocated amount. Always try to get the answers from the participants.

Figure 8.3   Chair game

Source: Alberta Council for Global Co-operation.

Explain that the ten people represent the world's population, which is roughly 6 billion – so each person represents 10 per cent or 600 million people.

Question: What percentage of the world's population is in G8 countries?

Answer: 12 per cent.

Nominate one person (ideally at one end of the line) to represent the G8. The remaining 88 per cent are the majority world.

Question: What percentage of the world's total Gross Economic Output is produced by the G8 countries?

Answer: 48 per cent (roughly 50 per cent).

Ask the nominated G8 person to occupy five chairs while the remaining nine squeeze onto the other five chairs.

Question: What percentage of the world's total annual carbon emissions are produced by the G8 countries?

Answer: 62 per cent.

Ask the nominated G8 person to occupy six chairs while the remaining nine squeeze onto the other four chairs.

Question: Of the top 100 multinationals how many have their headquarters in G8 countries?

Answer: 98 per cent.

That would leave the majority without any chairs but if the G8 generously gave a bit of aid that would leave them with one chair. Ask the nominated G8 person to occupy nine chairs while the remaining nine squeeze onto only one chair. This is obviously quite difficult.

Outcomes and tips: Ask the G8 how he/she is feeling – then ask the majority world what they would do to change the situation. Some people might try to persuade the G8 to give them their chairs back; others just go and take them.

4. Debate it!
Often participants really value an opportunity to talk freely, but as a facilitator free debate can be very difficult to structure and dominant personalities or viewpoints can easily take over. The following activities help to structure things.

The YES/NO game (or Issue Lines)
Aim: To see opinions in relation to other points of view and for participants to try and defend or persuade others of their perspective.

Method: All stand up and explain that you are all standing on a long line with YES at one end and NO at the other and NOT SURE somewhere in the middle. (It can help to make signs.) Read a statement and ask people to position themselves in the room depending on their point of view. When people have moved, ask someone standing in the YES or NO sections to try and explain why they are standing where they are, then ask the opposite side for an opinion. Allow the debate to continue awhile and then ask participants to reposition themselves depending on what they have heard.

Questions that we have used include:

★ Is nuclear power a viable alternative to fossil fuels?
★ Is it desirable that levels of consumption in the developing world equal that of those in the 'developed' world?

Outcomes and tips: These debates are often lively. Be careful not to allow any one person, including the facilitator, to dominate and make sure the question has a possible yes–no range of answers. Try working in smaller groups to allow everyone to speak and then give time for groups to feed back their main points.

Role plays

Aim: To present different opinions and encourage people to think from varied perspectives. Role plays enable participants to develop characters and take on their opinions, providing an excellent opportunity to express common misconceptions and controversial opinions without the participants speaking personally.

Method: Prepare a 'pro', 'anti' and/or 'neutral' camp with prompt cards for each. This should include context, details on how to act and speak, and ideas on how to respond to questioning. Explain that people should keep in this role at all times, even if they don't agree with the views expressed. Give people time to discuss and expand on the prompt cards. A good way to structure discussion is to chair a hearing between the different parties, where a mediator asks each side to present their case in turn. Allow time for open questions, followed by a summing up.

Example: Building a local road.

★ Chief Executive of Gotham City: Your city is booming and the key to its success is road transport. Business and tourists are being attracted from the whole country. Argue that: if the new ring road doesn't go ahead then the economic viability of the area will suffer. Less growth means fewer taxes, which means less money for public services.
★ Concerned citizens near the proposed road: There has been so much development in this city that there doesn't seem a case for any more. Roads are jammed already and just building more roads doesn't solve the problem, but only encourages more car use. Argue that: more cars equals more pollution, accidents and unhealthy lifestyles.

Outcomes and tips: Role plays need to be well prepared and work best when people are confident speaking in front of each other. With a longer session, ask participants to research and develop roles for small groups to enact.

5. Connecting histories and lives

Sharing our collective pasts is a key way to begin to understand our present and to imagine our futures. There are many ways to do this, through oral histories, participatory video documentaries, etc. It can also be useful to plot events onto a visual representation of history.

The rise of global capitalism and resistance timeline

Aim: To chart the rise of the current economic system and global resistance to it, to show international organisations in context. It can be used, for example, to show how US foreign policy has worked and evolved, or how resistance movements in the global North and South have progressed and connected.

Method: Draw a timeline on a big piece of paper or cloth, write key moments of the development of the economy and resistance events on to cards and give one or two to each pair. Give them time to discuss it and ask any questions about it and then ask them to put the events on the timeline where they think it occurred. Also give participants blank cards and ask them to fill in things they would like to add – maybe from their local area or that they have been inspired by. Go through the events, asking others to explain and give their opinions and help people identify connections.

Outcomes and tips: This activity helps people see connections between seemingly separate events. Make sure you have reliable information on dates, etc.

6. Get out of the classroom – creative educational events

Plays, film screenings, music, talent shows, bike rides, mural painting, nature trails, and cooking are all ways to get together and can be adapted to a theme. A walking tour can be a great way to bring a theme to life and to learn about our built environment or local history.

Walking tours of immigration controls and the 'chain of deportation'

Aim: To draw to people's attention institutions, companies and government departments involved in the chain of deportation of asylum seekers. A tour exposes the process and joins the dots in the picture of detention and deportation and helps understand the system.

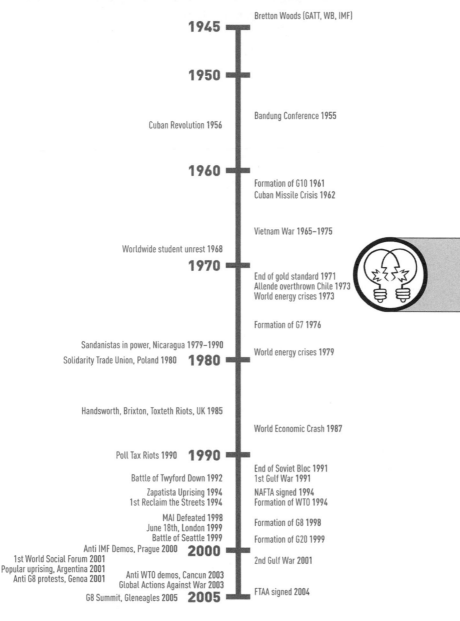

Figure 8.4   Timeline of world history and resistance used in a workshop

Source: Trapese Collective.

Method: Research the government sponsored institutions and private companies that earn money carrying out these racist policies, and organise a tour to visit some of the places in your area that are involved in locking up and deporting asylum seekers. Go as a walking tour in groups, assemble in a public place and have easily identified guides with maps, information, loud speakers, music, etc.

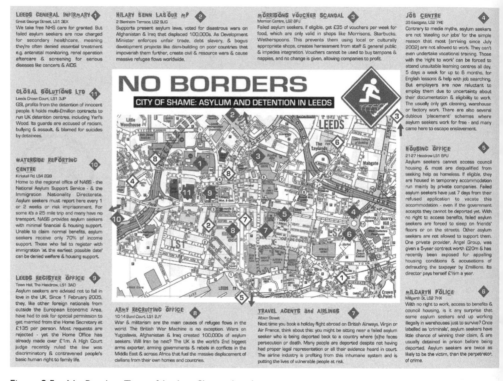

Figure 8.5   No Borders Tour of Asylum Shame, Leeds

Source: Leeds No Borders.

Social change pub quiz

Aim: A social event where the content matter is related to important issues. Can be a good way to outreach to different audiences.

Method: Find a venue to host you – community centre, student union bar or a local pub – and advertise the event. Make up several themed rounds of questions, get answer sheets, pens and prizes and maybe a microphone. Keep it varied by using multiple choice, picture or music rounds, bingo or maybe even some subverted karaoke.

Example: The Food Round.

1. How many million battery chickens are produced for consumption each year in the UK?

    (Answer: 750 million)

2. How many billion did Wal-Mart's global sales amount to in 2002; was it (a) $128 billion, (b) $244.5 billion, or (c) $49 billion?

    (Answer: b)

3 What is the legal UK limit for the number of pus cells per litre of milk that may be legally sold for human consumption?

    (Answer: Up to 400 million)

Sources:
1. and 2. Corporate Watch (2001)
3. Butler (2006).

7. Planning for Action

The aim of the game is that people leave with concrete ideas about what they are going to actually do – a next date, an ambition or a vision. This stage is also a chance for people to share ideas about the things they are already doing and plug any events or projects.

Action mapping

Aim: To show a variety of actions and inter-connections.

Method: Ask groups to think of two ways to tackle the issue that the workshop is dealing with at different levels: the individual, local and national/international for example. (See Table 8.1 *overleaf*.)

Outcomes and tips: Get groups to think about time scales for their actions and how they practically might do them. A variation is to think about two things to do this week, this month, this year, etc.

Picture sequences

Aim: To look at how things are and how we would like them to be, and to work out how to get there.

Method: Draw a simple picture that represents, 'the present', with all the problems illustrated. Then, as a group, put together a second drawing to represent 'the future', which shows the same situation once the problems have been overcome or the improvements made. Make sure you incorporate everyone's ideas of what you hope

Table 8.1   *Example*: The problem: climate change

What can you do individually?	What can you do locally?	What can you do nationally/ internationally?
Energy efficiency; insulation, turn heating down etc., switch to green energy sources.	Develop community owned renewables, energy sources and food production.	Take part in actions, protests, camps and gatherings.
Stop flying unless unavoidable and cut car use.	Set up neighbourhood composting schemes.	Support/publicise struggles against fossil fuel extraction, e.g. new pipelines.

to achieve. Once you have the drawings, put them where everyone can see them with a space in between them, then ask yourselves how you can get from the first to the second. What needs to happen to get there? How could it be achieved? Use the answers to make your own middle, between the present and the future. Now you have your vision and have gone some way to working out your action plan.

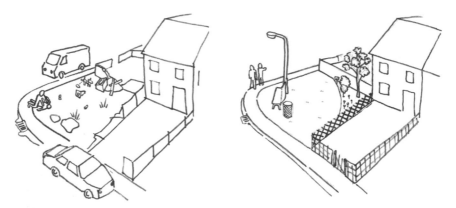

Figure 8.6   Designing your own life: before and after

Source: Guy Pickford, adapted from Groundswell Toolkit for Change.

Presents

Aim: To end the workshop on a high note and to get participants to 'think the impossible'.

Method: Identify the main problem that people want to focus on. Give out cards with an imaginary present written on to each participant. Ask them to describe how they would use their present to solve the problem.

Examples:

★ the ability to look like anyone you want
★ £1 million
★ a minute of primetime TV
★ a key that unlocks any lock
★ an invisibility cloak
★ a guarantee you'll never get caught.

Calendars of resistance   It's important to share information on other things that you know are going on in the area. Draw up a calendar with contact details for people to get more information.

Skills for good communication

★ Challenge dominance. Both from vocal participants and as facilitators. Be open from the start about why activities are being undertaken and do not manipulate participants to certain ideological ends.
★ Don't judge. Be supportive in your approach and recognise the validity of a diversity of actions and viewpoints. It's not about persuading people to think or act as you want them to!
★ Listening is crucial. Learn the importance of active listening to allow necessary discussion. Letting people talk, reducing dependency and empowering people to think for themselves are at the heart of radical education.
★ Overcome powerlessness. 'If everything is connected you can't change anything without changing everything. But you can't change everything, so that means you can't change anything!' (A student after a lesson on globalisation from the book *Rethinking Globalisation* (Bigelow and Peterson 2003)).

While it's not true that we cannot change anything, this student's comment demonstrates how depression and a feeling of powerlessness is a logical reaction when solutions seem very small in the face of such large forces. Here are some tips for giving positive workshops about negative subjects:

★ Don't cram in too much information. Go step by step and give things time and space to develop.
★ Mention the empowering side. Look for the positive things we can do and emphasise our creativity, our adaptability. Sharing personal experiences and failures can be very useful. Start with concrete achievable aims and develop from there.

☆ It is important that people are not left feeling isolated and that there is follow-up.

☆ Try making a presentation about initiatives or protests that have inspired you with images or photos and use this is a springboard for talking about the viability of these ideas.

making the leap to action

There are lots of issues which people are angry and passionate about and areas where they want to take action. Getting together to discuss and understand problems is a good way to reduce feelings of isolation and to launch campaigns and projects. It is really important to pick our starting points carefully, to build up trust and meet people in their daily realities, whilst not being scared of expressing radical views. Having worked with different groups we have been continually inspired by people's views, opinions and desires to instigate change. These experiences have helped us to break down the false distinction between activists and everyone else and we have learned as much as we have taught. Popular education is about building from the beginning and finding innovative ways to learn together, realising the capacity that we have to take control of our lives and facilitating collective action, and for us, this lies at the heart of building movements for change.

The Trapese Popular Education Collective is based in the UK and since 2004 has been working with groups of adults and young people to understand and take action on issues including climate change, globalisation and migration. They also produce educational resources and promote participatory, interactive learning through training and skill-shares (see www. trapese.org). Additional material sourced gratefully from Rising Tide, Groundswell, the Alberta Council for Global Co-operation and Rethinking Schools.

resources

Books

General

Acklesberg, Martha (1991). *Free Women of Spain, Anarchism and the Struggle for the Emancipation of Women*. Edinburgh: AK Press.

Bigelow, B. and B. Peterson (2003). *Rethinking Globalisation*. Rethinking Schools Press, USA.

Buckman, P. et al. (1973). *Education Without Schools*. London: Souvenir Press.

Butler, J. (2006). *White Lies – Why You Don't Need Dairy*. London: The Vegetarian and Vegan Foundation.

Carnie, F. (2003). *Alternative Approaches to Education: A Guide for Parents and Teachers*. London: Routledge.

Corporate Watch (2001). *What's Wrong with Supermarkets*. Oxford: Corporate Watch.

Crowther, J., I. Martin and M. Shaw (1999). *Popular Education and Social Movements in Scotland Today*. Leicester: NIACE.

Crowther, J., I. Martin and V. Galloway (2005). *Popular Education. Engaging the Academy. International Perspectives*. Leicester: NIACE.

Katz-Fishman, W. and J. Scott (2003). *Building a Movement in a Growing Police State. The Roots of Terror Toolkit*, www.projectsouth.org/resources/rot2.htm

Freire, P. (1979). *Pedagogy of the Oppressed*. London: Penguin.

Freire, P. (2004). *Pedagogy of Indignation*. London: Paradigm.

Giroux, H.A. (1997). *Pedagogy and the Politics of Hope. Theory, Culture, and Schooling: A Critical Reader*. Colorado: Westview Press.

Goldman, E. (1969). 'Francisco Ferrer and the Modern School'. In *Anarchism and Other Essays*. London: Dover, 145–67.

Goodman, P. (1963). *Compulsory Miseducation*. New York: Penguin.

Gribble, D. (2004). *Lifelines*. London: Libertarian Education.

Hern, M. (2003). *Emergence of Compulsory Schooling and Anarchist Resistance*. Plainfield, Vermont: Institute for Social Ecology.

hooks, bell (2004). *Teaching Community. A Pedagogy of Hope*. London: Routledge.

Horton, M. and P. Freire (1990). *We Make the Road by Walking: Conversations on Education and Social Change*. Philadelphia: Temple University Press.

Illich, I. (1973). *Deschooling Society*. London: Marion Boyars.

Koetzsch, R. (1997). *A Handbook of Educational Alternatives*. Boston: Shambhala.

Muttitt, G. (2003). *Degrees of Capture. Universities, the Oil Industry and Climate Change*. London: Corporate Watch, Platform and New Economics Foundation.

Newman, M. (2006). *Teaching Defiance: Stories and Strategies for Activist Educators*. London: Wiley.

Ward, C. (1995). *Talking Schools*. London: Freedom Press.

Free schools

Gribble, D. (1998). *Real Education. Varieties of Freedom*. Bristol: Libertarian Education. .

Novak, M. (1975). *Living and Learning in the Free School*. Toronto: Mclelland and Stewart Carleton Library.

Richmond, K. (1973). *The Free School*. London: Methuen.

Shotton J. (1993). *No Master High or Low. Libertarian Education and Schooling 1890–1990*. Bristol: Libertarian Education.

Skidelsky, R. (1970). *English Progressive Schools*. London: Penguin.

Popular histories

Hill, C. (1972). *The World Turned Upside Down.* London: Maurice Temple Smith.
Morton, A.L. (2003). *The People's History of England.* London: Lawrence and Wishart.
Thompson, E.P. (1980). *The Making of the English Working Class.* London: Penguin.
Zinn, H. (2003). *A People's History of the USA.* London: Pearson.

Websites

Catalyst Centre www.catalystcentre.ca/
Centre for Pop Ed www.cpe.uts.edu.au/
Development Education Asociation www.dea.org.uk/
2002 Education Facilitators Pack www.web.ca/acgc/issues/g8
Education Otherwise www.education-otherwise.org/
Highlander School www.highlandercenter.org/r-b-popular-ed.asp
Home Education www.home-education.org.uk
Institute for Social Ecology www.social-ecology.org/
Interactive Tool Kit www.openconcept.ca/mike/
Intro to Paulo Freire www.infed.org/thinkers/et-freir.htm
Intro to Pop Ed www.infed.org/biblio/b-poped.htm
Laboratory of Collective Ideas (Spanish) www.labid.org/
PoEd News www.popednews.org/
Popular Education European Network List http://lists.riseup.net/www/info/poped
Popular Education for Human Rights www.hrea.org/pubs/
Project South www.projectsouth.org/
Trapese Collective www.trapese.org

Film resources

Beyond TV www.beyondtv.org/
Big Noise Films www.bignoisefilms.com/
Carbon Trade Watch www.tni.org/ctw/
Eyes on IFIs www.ifiwatchnet.org/eyes/index.shtml
Global Exchange http://store.gxonlinestore.org/films.html
Undercurrents www.undercurrents.org/

9 why we are what we eat

Alice Cutler and Kim Bryan

Food is essential to what we are. For centuries it has shaped societies and cultures. The word 'diet' stems from the Greek, *digitals*, which means 'way of life'. However, huge changes are underway in our ways of life through the corporate takeover of the food chain whereby food is placed increasingly in the hands of multinationals and locked in a cycle of fossil fuel politics. In the last 60 years how we eat, produce, consume and cook food has altered dramatically. From pumping it with additives and preservatives, growing it with pesticides and fertilisers, to the consumption of processed and genetically modified foods, a chain of events has been set in place by industrial agriculture that has huge implications for farmers, the countryside and biodiversity in general.

A crucial part of building sustainable futures, self-managed lives and struggles for autonomy lies in addressing how we produce what we eat. The premise of this chapter is that the issue of food and food production is inherently political. To take a step away from the grip of capitalism and in order to free ourselves from being passive consumers who are alienated from nature, we must learn how to nourish ourselves and our movements. This chapter explores a range of possibilities and projects that lend themselves both as inspirational examples and real and viable options in changing how and what we eat.

understanding where we are at

Imagine the history of humankind stretched out along a 10-metre rope. The blip of time since the Industrial Revolution would be represented by the last few centimetres on the rope with industrial agriculture and genetic modification of crops barely measuring a few millimetres at the end. The first farms began to appear as plant and animal species were domesticated about 13,000 years ago but it wasn't until the industrial revolutions began to unfold in eighteenth-century Europe that mass movement from the country to the city took hold and the way food was produced changed significantly. In the UK the end of the Corn Laws in 1846 saw the end of market protectionism and ever since imports have increased. By the start of World War II, 70 per cent of the UK's food was imported. This trend is replicated around the world as free market policies have radically changed traditional agricultural models. The following sections look at three of these issues in more detail: corporate control, the environment and health.

Corporate control

There are many negative, hidden effects to the amazing variety of cheap products that the world food trade brings consumers in the rich North. Food is big business: the global trade is estimated to be worth $4 trillion dollars a year and the market is concentrated in the hands of a few powerful companies. This imbalance has been created over the last 30 years through food and agriculture policies as well as global trade agreements which promote trade liberalisation and the globalisation of the food economy. The quality and nutritional value of food, the livelihoods of small-scale farmers, producers, agricultural workers and community economies are all secondary concerns to the profit margin in the industrial agricultural model. The European Common Agricultural Policy (CAP) has overseen the creation of a heavily subsidised agriculture system, consisting of ever larger farms which drive down production costs through economies of scale and technical efficiency. Meanwhile, World Trade Organisation (WTO) policies have aggrieved the economic crisis in rural areas throughout the world. The liberalisation of agricultural markets, the forced opening of borders and tariff cuts have put farmers in a global system of unlimited competition in which the main beneficiaries are transnational agribusiness companies and their shareholders. The interests of small producers and farmers are ignored causing devastating effects on the livelihoods of billions.

> My warning goes out to all citizens that human beings are in an endangered situation. That uncontrolled multinational corporations and a small number of big WTO members are leading an undesirable globalisation that is inhumane, environmentally degrading, farmer-killing, and undemocratic. It should be stopped immediately. Otherwise the false logic of neoliberalism will wipe out the diversity of global agriculture and be disastrous to all human beings. (Lee Kyung Hae, South Korean farmer who killed himself at the WTO protests in Cancun in protest against unfair trade subsidies)

The tragic effect on farmers across the world is one part of the untold damage of modern day agriculture. Supermarkets have increasingly eroded local choice as smaller, independent shops struggle to compete. Wal-Mart in the US (whose UK subsidiary is Asda) is the biggest retailer in the world. Thousands of stores, including newsagents, post offices, grocers, bakers and butchers, have closed creating virtual ghost towns as the number of large out-of-town stores increases. These all-powerful companies dictate the terms and conditions of production to millions of small farmers and suppliers who are forced to compete for a limited number of agreements to supply. For example, many orchards in the UK have been abandoned simply because the varieties of apples produced do not transport or keep well and are therefore not financially viable for supermarkets to stock. Over the last 30 years, 60 per cent of orchards in the UK have been destroyed. By 1996 the UK imported 434,000 tonnes of apples, nearly half from outside Europe.

Corporate control of food produce doesn't just stop at the supermarket or production line. Just three corporations control one quarter of the world's seed market (Monsanto, Syngenta and DuPont) and biodiversity is not high on their agenda. National seed lists in many countries make it illegal to buy and sell unusual varieties and it is prohibitively expensive to keep seeds on the list. Agribusi- nesses require farmers buying seed to sign contracts that prevent them saving and replanting seeds at a later date. As this is difficult to enforce, seeds are now being genetically modified to be sterile after a year in order to protect company's patents, a process known as terminator technology. While this guarantees profits, an estimated 1.4 billion farmers worldwide depend on seeds saved or exchanged with neighbours. The Chilean Rural and Indigenous Network call this copyrighting or patenting of living things 'a crime against humanity'.

Environmental impacts

The food system has become a major contributor to climate change. The emissions that a typical UK family of four are responsible for each year equals 4.2 tonnes of CO_2 from their house, 4.4 tonnes from their car, and 8 tonnes from the production, processing, packaging and distribution of the food they eat (Sustain/Elm Farm Research Centre Report 2001). As the world heads towards the peaking of oil supply, the extent that our food systems depend on fossil fuel energy will be brought in to sharp focus. Between 1950 and 1990 the world population doubled and in large part it was the much greater use of chemical (oil derived) fertilisers that allowed this. Many millions of farmers in the global south were devastated by increases in oil prices in the 1970s after the 'green revolution' had dramatically changed the way that they grew food through a more industrial model based upon petrochemical fertilisers and pesticides. In 1940, the average farm in the USA produced 2.3 calories of food energy for every calorie of fossil energy it used. By 1974 (the last year in which anyone looked closely at this issue), that ratio was 1:1. Now it is much higher. For example 66 units of energy are consumed when flying 1 unit of carrot energy from South Africa (Manning 2004a).

Meat production systems are very energy and water intensive. Producing a kilo of beef, for example, uses three times the food energy it yields. The methane from global dairy herds has become a major source of human-induced greenhouse gas emissions. Methane is a by-product of digestion for cows, sheep, goats and other livestock but it is also a potent greenhouse gas – over ten times more powerful than CO_2 over a 100- year period in terms of its 'greenhouse' effect. Global annual methane emissions from domesticated animals are thought to be about 100 million tonnes or about 15 per cent of the annual methane emissions from human activity, the others being production of fossil fuels, wet rice cultivation, biomass burning, landfills and domestic sewage. Scientists believe that current methane emissions will account for over 15 per cent of human-induced climate change over a 100-year time frame (IUCC 1993). It takes almost 800 kg of cereals (used as feed for animals) to produce the meat an average North American eats in a year. This is nearly five times more grain that a reasonably well fed African eats in a year. Arguably a switch to a vegetarian or vegan diet can do more to cut an individual's carbon footprint than many other measures.

The distances that food travels is a major issue. It can be cheaper to buy at a distant location and bring the food in (despite the cost of freight and shipping) rather than buy from local suppliers. One consequence of this is that a large proportion of road freight moves food around very long supply lines. Between 1989 and 1999 there was a 90 per cent increase in road freight movements of agricultural and food products between the UK and Europe (see DETR 2000). Rather than importing what they cannot produce themselves, many countries appear to be simply 'swapping food'. In 1997, the UK imported 126 million litres of milk and exported 270 million litres (FAO Food Balance Sheet Database 2001, see www.fao.org). One study has estimated that UK imports of food products and animal feed involved transportation by sea, air and road amounting to over 83 billion tonne-kilometres, using 1 billion litres of fuel and resulting in 4.1 million tonnes of carbon dioxide emissions (Shrybman 2000). Once inside the UK, food continues to clock up food miles, indeed food, drink and feed transportation accounts for up to 40 per cent of all UK road freight (Jones 1999).

Air freighted fruit and vegetables are a particularly fuel intensive and unsustainable trend which has impacts that go far beyond the environment. Millions of workers in the global South are at the mercy of supermarket production schedules. Orders are dependent on the previous day's sales, and so already precariously employed workers' lives are dictated by volatile production schedules. There are widespread allegations of worker rights abuses (see for example the Ethical Trading Initiative report, 2003). In Kenya, women work day and night in refrigerated packing sheds producing high value goods such as bundles of asparagus shoots, miniature corn, dwarf carrots and premature leeks, tied together with a single chive. The chives and plastic trays are flown out from England to Kenya where the produce is prepared at low labour costs. They are then air-freighted back to England again, a round trip of 8500 miles (Growers' Market, Felicity Lawrence, *Guardian*, 17 May 2003).

Waste is another key environmental issue related to food. In addition to very wasteful use of oil for excessive packaging, rotting food in landfill sites produces methane, a potent greenhouse gas and leachate which can contaminate groundwater supplies. Whilst around 40 per cent of waste produced in the average UK home is compostable, only 3 per cent is composted. This adds to the estimated 30 per cent of perfectly edible food a year that ends up in landfill.

Health

Malnutrition, starvation and famine remain huge issues in the twenty-first century. Conflicts and climate change are devastating small-scale food production in the global South. According to Oxfam, the number of food emergencies in Africa has nearly tripled since the mid 1980s. Meanwhile, huge tracts of land in the so- called developing world are used to produce animal feed or other crops for export. Hunger, of course, is not caused by lack of food, but by grossly unequal access to the plentiful food that exists. In the USA, there are an estimated 10 million low-income Americans who do not have enough to eat. People on low incomes are more likely to suffer and die from diet related diseases such as cancer and coronary heart disease (Food Poverty Project 2002). To get all the calories needed in a day while spending the least money, the best bet is to go for a high fat, sugary diet and avoid fresh salads and fruit, a situation which has widespread impacts on health.

A range of other health issues, from allergies to behavioural problems in adolescents and Alzheimer's disease, have been linked to food. Non-organic vegetable produce not only has less nutritional value but it is laden with chemicals and hormones. A number of substances found in everyday foods are carcinogenic. For example, aspartame, a sweetener found in many foods, has also been linked to leukaemia and lymphoma cancer, yet there is very little research carried out into these everyday foods before they are placed on the market. Every so often a food scandal or crisis erupts such as

BSE, foot-and-mouth disease or salmonella, which hints at the potential time bomb waiting to explode related to the long-term health effects of the cocktail of chemicals and drugs we consume.

Box 9.1   Hartcliffe Health and Environment Action Group (HHEAG)

In response to a damning report investigating health in Bristol, UK, a group of local people set up a project to support residents in taking control of their local environment, food and health. They raised funds to install new school kitchens, and doctors can now refer people for classes on nutrition and cooking. The project also has two gardens – one open to anyone interested in learning about or growing food and another market garden with paid employees that supplies the food co-op shop.

resistance is fertile

Developing food sovereignty and taking back control of what we eat have become vital and important routes for action. Across the world, movements in urban and rural areas are fighting for the rights to their land and for food sovereignty. Although there would inevitably be less consumer choice than the artificially high levels we have today, a different food production system which prioritises short supply routes could massively cut greenhouse gas emissions, and is essential in the battle against climate change. Beyond this, if we stop buying these environmentally, socially and personally damaging foods and instead set up co-operatives, 'grow our own', build community gardens and support local farmers, then the economic, agribusiness system is starved of nourishment and weakened. Many of the ideas in this chapter are not new, indeed we don't have to look that far back into the past to see pointers for the future. Up until the middle of the last century the practices of seed saving and swapping and organic agriculture were all commonplace.

The following sections look at four ways that people are taking back control over the food that they eat and the way it is produced: community food projects, sustainable agriculture, food co-operatives and movements for change.

Figure 9.1   Resistance is fertile. Brazil: Indigenous people reclaiming their traditional lands and cutting down eucalyptus trees owned by Aracruz Cellulose

Source: Carbon Tradewatch.

Box 9.2   *Fighting for food sovereignty: Via Campesina*

Via Campesina is an international movement, which co-ordinates organisations of small and medium-sized producers, agricultural workers, rural women and indigenous communities. It is autonomous from all political and economic organisations. Since 1992, peasant and farm leaders have organised under this banner around issues of agrarian reform, credit and external debt, technology, women's participation and rural development. Mass protests and land occupations have brought attention to destructive free trade agreements and the structural violence of corporations that transform the land into a commodity and dump subsidised foods on the markets of poor nations. These movements are not only an inspiration but also an important reminder of the need for solidarity with global struggles to challenge international institutions, such as the WTO.

Fighting for food sovereignty

The concept of food sovereignty was a term coined by Via Campesina at the 1996 World Food Summit in Rome and is defined as the right of peoples, communities and countries to define their own agricultural, pastoral, labour, fishing, food and land policies which are ecologically, socially, economically and culturally appropriate to their unique circumstances.

Growing communities: community food projects

Reclaiming land as a community resource has been part of an ongoing struggle around the world. There are many wonderful examples of community gardens, emphasising education, the production of healthy organic foods and also providing a valuable base for community interaction. Community gardens not only fulfil social needs but can also relate to the political formation of movements for social change. In the 1970s the fiscal crisis that gripped the USA impacted heavily on inner-city areas. In New York, dismayed by government inaction and the increasing number of vacant lots, crumbling buildings and rubbish-strewn streets, a group of people known as the 'green guerrillas' began to set up community gardens. By the early 1990s the network of 850 community gardens on abandoned plots of land in New York had become an important social resource, providing a place to meet and talk, repair broken bikes, play music, and grow food and herbal medicines.

> The gardens became catalysts for community development. Once people succeeded with the garden, they went on to other things like fixing the schools, housing, creating jobs, whatever was needed. (Ferguson, 1999)

However by the late 1990s, the then mayor, Rudi Giuliani, began to pursue an aggressive policy of gentrification in inner-city areas. A series of high-profile demonstrations and campaigns took place to resist evictions, but today only 50 community gardens remain in New York. However, the idea of guerrilla gardening has spread across the world. Your imagination is the only limitation in looking for places to grow food – railway embankments, back gardens, golf courses, roofs, car parks, overgrown areas and cracks in the pavement can all become areas to grow edible crops.

Community gardens show that urban food production is possible, but making food accessible to people on a wide scale is a major challenge. Price and convenience are often the major determinants of what people buy and many areas are fresh food deserts with no greengrocers. Growing at home in window boxes, containers or small gardens is a vital response. A very small area can supply fresh leaves, beans

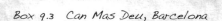

Box 9.3 Can Mas Deu, Barcelona

Can Mas Deu is a squatted social centre, home to around 25 people on the outskirts of Barcelona. Since 2001 the old hospital building's finca (estate) which had stood empty for over 50 years, has been reclaimed, the ancient irrigation system has been restored and now the terraces are bursting with vegetables, herbs and flowers. As well as providing food for the house there are over 100 people who are involved in the Horts Comunitaris or Community Gardens. Notices around the neighbourhood invited anyone to take on a plot, provided they were interested in cultivating food organically and prepared to embark on a journey of collective work, learning and skill sharing. There has been a waiting list for plots ever since. It has been an inspiring experience, not only for the opportunity to learn about growing food but also the interchange of knowledge and skills between generations and nationalities.

Figure 9.2   The Cre8 Summit Community Garden, Glasgow, located on the site of the proposed M74 motorway extension

Source: Cre8 Summit.

and potatoes. For many people gardens are a symbol of self-reliance rather than dependence on a system to which they have no control. 'Ecological footprint' refers to the amount of land and water area a human would hypothetically need to provide the resources required to support itself and to absorb its wastes. Footprinting is used around the globe to measure the environmental sustainability of an individual, organisation or entire nations (see, for example, www.ecofoot.org). If we are truly to reduce our ecological footprints, then we will have to redesign both our urban and rural spaces to maximise food production. Below the concrete there is a garden!

Sustainable and organic agriculture

Taking back control of our lives involves moving rapidly away from industrial agricultural models and towards sustainable farming practices. Sustainable agriculture is farming that maintains the natural environment and sustains resources whilst respecting all involved, from farm workers and consumers to the animals raised for food. Despite the onslaught of intensive agricultural models some traditional farming practices have survived. The last decade has seen a resurgence in interest from both consumers and farmers and an increase in research and experimentation in sustainable farming practices. Sustainable agriculture is characterised by its emphasis on crop diversity and rotation, favouring small and medium-sized farms and prioritising staple crops rather than cash crops for export. It uses natural systems and cycles to grow crops, improve yields and reduce the needs for pesticides and chemical inputs. Organic agriculture also cuts the amount of oil energy in crops as long as they are produced locally.

Box 9.4 Sustainability Cuban style: Los Jardineros Urbanos (Urban Gardens)

In the residential barrio of Miramar in Havana, hidden amongst the concrete, is a patch of land lush with vibrant green. The urban garden, one of the first in Havana, started 15 years ago and was given by the government to support the local community. The 2–3 acre site includes a polytunnel, community shop and small research unit. The vegetable beds are arranged in raised strips and full of flourishing vegetables, fruits, flowers, house plants and herbs. At regular intervals along the rows, metal plates are covered in liquid solutions to attract certain insects and deter other pests that might damage the crops. The gardens produce lettuce, cabbages, herbs and peppers all the year round as well as more seasonal vegetables which are given to the local schools and sold in the community shop. Fifty per cent of the profits go to the state and the workers in the garden and shop take the rest. Volunteers come from all over the world to work in this garden as well as Cubans who come for advice on setting up their own community gardens.

Cuba is an enlightening example of a country that has fully embraced sustainability in food production. In the early 1990s when the Soviet Union collapsed, Cuba's financial support dried up and the country was faced with widespread shortages. With few options to import food given the stringency of the US embargo, Cuba converted almost entirely to an organic, non-oil dependent, production system within ten years. The methods have borne some amazing results, not only in terms of food production but also in the development of a more personalised food culture, woven deeply into patterns of food consumption, nutrition and community. Farmers and urban citizens dedicated themselves to meeting food demands and urban plots and parcels of land that had formerly operated as cane plantations for the sugar industry were turned over for domestic food production. In 2002, Cuba produced 3.2 million tonnes of food on more than 29,000 urban farms and gardens (Barclay 2003).

Alternative agricultural systems can create more equal relationships between the environment, community and producers. An example which has blossomed over the last decade are Community Supported Agriculture Schemes (CSAs). It is a tall order for everybody to be active in their food production and so CSAs act as a bridge between community and sustainable agriculture. For example, Stroud Community Agriculture in the UK employ the equivalent of one full-time farmer and the 50 members collect vegetables each week and take a share of the meat. Importantly, the consumers share the risk, their regular contributions pay the wages of farm workers, and the yields are shared out equally.

Food co-operatives

Going beyond vegetable produce, to drive down costs of ethically sourced staples, such as rice and pulses, and to step away from dependence on supermarkets many people turn to food consumer co-operatives. These are worker or customer owned businesses that provide high quality and value food to their members. Robert Owen, Welsh industrialist and socialist and one of the founders of the co-operative movement, believed in putting his workers in a high quality environment with access to education for themselves and their children. He had the idea of forming 'villages of co-operation' where workers would pull themselves out of poverty by growing their own food, making their own clothes and ultimately becoming self-governing. Food co-operatives are one way of self-organising that have become widespread, along with housing, workers, renewable energy, social care, banking and agricultural co-operatives.

There are a number of benefits to setting up a food co-operative: they can reduce food costs, improve nutrition and allow members to negotiate directly with growers or producers. Large wholefood, wholesale co-operatives supply smaller local food co-operatives and members decide on the pricing of the goods. The 'Fruity Nutters' co-operative in the UK add 5p to organic products, but 5 per cent to non-organic goods, thus subsidising organic foods and making them more accessible to people on lower incomes. Any profit from the co-operative covers expenses, and any surplus goes to good causes. Co-operatives can achieve lower prices than supermarkets, not only by buying direct, but also by dividing up bulk orders themselves, thus avoiding the financial and environmental costs of excessive amounts of packaging. Food co-operatives have enormous potential for expansion. In Japan, one in five people are part of a consumer co-operative. In the UK in the year 2004–05 UK health food wholesalers and shops reported a rise of 35 per cent in sales, indicative of an increased awareness of health and environmental concerns (Union of Co-operative Enterprises, www.uk.coop).

Nourishing movements for change

The formation of cooking collectives and community kitchens is another way that food consumption can be more sustainable and sociable. Cooking collectively at social centres, protest camps, mobilisations and gatherings is often done by grassroots, non-profit-orientated groups. Unlike commercial caterers they are based on the spirit of mutual aid where people help with chopping vegetables and washing up. There are numerous kitchen collectives across the world that run on these principles. The Dutch collective Rampenplan, for example, have been running for over 20 years and can provide locally sourced organic food for up to 5000 people. Other collectives include the Anarchist Teapot who, as well as cooking for events, help to run a permanent cafe

Box 9.5   Food Not Bombs

Food Not Bombs is a loose network that started in the USA which recycles large quantities of food, cooks it up and distributes it for free. It is a radical political act in today's society to distribute this food and as a result, Food Not Bombs groups have met with massive police repression. By the year 2000 there had been 1000 arrests in San Francisco alone, vehicles were impounded, people were beaten, detained and even jailed for participating in Food Not Bombs. However, in 2000 there were 175 Food Not Bombs groups in the world, distributing food with no strings attached but 'as a celebration of life against death'.

four times a week at the Cowley Club in Brighton, UK, providing cheap and affordable food. These infrastructures are a valuable resource to weave food and consumption into our lives, and sharing food together is a time for talking and connecting with each other.

Freeganism   'We'll eat your scrap, but we won't buy your crap' (Freeganism slogan). The term 'freeganism' refers to a novel and widespread approach to delinking food consumption from corporate control. As one activist put it: 'It seeks to lessen rampant over consumption, environmental destruction, waste, and exploitation in the developed world.' Freegans get free food by pulling it out of the rubbish, a practice known as dumpster diving, skipping or recycling. Once you have got used to the idea of looking in bins freeganism allows you to avoid spending money on products that exploit the world's resources, contribute to urban sprawl, treat workers unfairly or disregard animal rights. Ample amounts of clean, edible food can be found in the bins of restaurants, supermarkets and other food-related industries and distributed to a wider net of people. The vast majority is still fine to eat, but has been thrown away due to strict hygiene and stock rotation laws.

facing up to the limitations

Access to land
The benefits of local, community based food production models are enormous but there are of course limitations. Small-scale growing projects demand a sustained amount of time, hard work and initiative from a group of committed people. Finding land on which to grow food can also be a major challenge due to a historic concentration of ownership, restrictive planning laws and the fact that agricultural buildings and land are increasingly more valuable when sold off as second homes or development land than as small-scale farms. Because local production works on completely different principles and strategies than the heavily subsidised economies of scale of industrial agriculture, competing with the international financial food market is virtually impossible. However, as the world moves towards peak oil scenarios, coupled with growing consumer demand and a rejection of the corporate control of food, the possibility of small-scale production is greatly enhanced.

Changing cultures
As with all topics in this book, the question of changing our individual desires and expectations is a big stumbling block. Making things accessible and affordable are

the biggest hurdles towards changing what people eat. In addition, adapting our bodies and taste buds to a new diet of food fresh from the allotment, grains and pulses from the food co-operative or strange seasonal vegetables can take a while. Cheap but tasty processed foods are more socially acceptable, largely due to fat, sugar, salt and flavourings, and the fact that they are also intensively promoted by the manufacturers as desirable foods to eat. The key is shifting the culture of food away from a 'McDonald's culture', part of which is challenging the idea that children have an innate preference for chips. Despite some successes, like celebrity chefs who encourage schools to introduce cheap, healthy food, many schools remain locked into long purchasing contracts with large, distant food contractors, such as Scolarest in the UK, who have monopoly contracts to ship in processed food. Education about the impacts of the food we eat on the environment and our health is crucial to ensure that from an early age the important issues of eating habits and diet are addressed.

Turning ideas into reality

Often, our lack of skills can hold us back in implementing our ideas. In response to this, there are an increasing number of growing projects for children, courses in organic gardening, permaculture and sustainable land use. Learning to grow our own food is a never-ending process as we adapt to new conditions and incorporate new ideas. With co-operation, education and skill sharing enormously successful projects can be created which are nourishing and fulfilling.

ways forward

> We are planting the seeds of a society where ordinary people are in control of their land, their resources, their food and their decision making. (Flier handed out at the Guerrilla Gardening event, London's Parliament Square, May 2001)

In this late capitalist society that has brought so much financial wealth, we are witnessing the devastating impacts of the corporate control of foods and industrial agriculture. In the last decade there has been a wave of popular responses: growing your own food, farmers' markets, fair trade, co-operatives, seed saving projects, allotments and traditional organic farming practices are all experiencing a renaissance. But there is still a long way to go to make these accessible and relevant to everyone. Despite the enormity of the challenges, connecting food production and consumption together can strengthen, green and nourish our communities. Most importantly, we must remember that our current food system needs to be changed as it is vulnerable, unsustainable and inefficient as well as being directly linked to wars over oil and global exploitation. This chapter has aimed to bring some of these issues into focus and explore alternatives to destructive food practices. It is a call to act – because we are what we eat.

10 how to set up a community garden

Kim Bryan and Alice Cutler

This guide is for anyone, anywhere who has walked past a derelict bit of land or has seen an empty allotment and imagined ... what if it were full of life and activity? Planting a garden is a lot about dreaming, visioning and creating. It can also be a lot of work; sometimes things don't grow, someone pulls out the carrots or slugs eat all the cabbages. But when things do grow and you can eat the results, seeing places transformed is amazingly rewarding.

Community gardens are small plots of land used for growing food which are organised along collective lines, usually for the benefit of the community. They have a huge range of potentially beneficial functions. They can:

★ Provide fresh, organic vegetables, fruits and herbs on your doorstep (or down the road) offering health, environmental and social benefits.

★ Bring people together to work on something which teaches useful skills, keeps people fit and healthy, and puts them back in touch with natural cycles and seasons.

★ Be a positive, practical demonstration of more sustainable living and 'doing it ourselves'.

★ Turn around abandoned land and create a beautiful space which increases pride in the neighbourhood.

★ Provide a home for birds, insects, newts, frogs and other wildlife as well as a space for humans to enjoy them.

★ Lead to increased environmental awareness.

★ Benefit people with learning difficulties, the elderly or people with behavioural problems through therapeutic work, positive activities and sensory gardens.

★ Preserve local varieties and biodiversity.

★ Provide spaces for kids, workshops for arts and crafts, bike repairs, social events or just a nice space to sit and chat.

setting up a community garden

There are lots of ways to locate potential plots: speak to and visit existing garden projects, contact the allotment or green spaces officer at the council or check out the land registry to find out if there are any abandoned bits of land. There are many 'squatted' community gardens where disused land is transformed into an urban oasis. In Glasgow, the Cre8 garden is situated in the way of a proposed motorway development and is a positive, visible community based protest. Gardens can take a few years to establish themselves and many choose plots with longevity in mind. Check out the community garden and allotment resources list at the end of this chapter for details of support and potential funding sources.

Dreaming and scheming

A community garden means making lots of collective decisions, so establishing how you will organise is important. As well as deciding what you want to grow (vegetables, flowers, fruit trees) you also need to think about what your main aims are (to produce food, educational, a calm place for people to relax) and who the garden is for. Agree

Box 10.1 Moulsecoomb Forest Garden Community Food Project, Brighton

'We started in 1994, and were rent free for the first year because the plot had been derelict for 20 years. We began by clearing the area and digging ditches to stop the soil being washed away as it's a very steep plot, constantly adding compost and manure to try and improve the land. Over time we started regular "open to everyone, no gardening experience necessary workdays," and became a charity – putting on regular events. Now there are seven plots, organic gardening, forest gardening, and an outlawed vegetable garden where we grow weird and wonderful heritage varieties not on the national seed list such as the Cherokee trail of tears, a French climbing bean. We have left areas open for wildlife, picnics and baking potatoes. We identify all plants that we weed and have found that nearly 95 per cent of plants have some use for us, one meal we had recently contained 26 different types of plant!' (Warren Carter, Seedy Business) on whether all or only some of the produce and tools will be shared as all these decisions affect your design. Come up with a name and think of the things you will need to create the garden (funding, materials to build with, structures) and divide up tasks and responsibilities. Consider logistics such as how can people contact the group, how much money you have got, how much you will need and who will look after the finances.

Ways of getting people involved

Once a garden exists, local residents are likely to stop and ask what it's all about, but in order to engage the local community, open days are great ways to entice people.

- ★ Food: Invite people to a picnic or barbecue to be held on the land. It's a great way to meet people, use and appreciate the land and build a community sense.
- ★ Open to all 'no experience necessary' work days: There will be lots of work clearing overgrown brambles or rubbish, preparing the beds, planting seeds, building sheds, setting up watering systems and a whole host more. Work days are a great way to share skills and get a lot done. Also, try non-work days, where people can nose about without feeling obliged to grab a shovel, such as bug hunts for kids, Halloween parties, etc.
- ★ Give out excess produce: When there is a bumper crop of fresh organic vegetable sharing any excess with people who live nearby is a good way of letting them see what you are doing and getting them on side!

Up until recently, back garden vegetable patches and allotments were very common, and someone who has had 50 years experience growing vegetables can be a gold mine of information. It's worth approaching people that have lived in the area for a while. They might know what the land was used for, soil type, local weather and what grows well there.

Design ideas

One of the keys to success with gardening is knowing the lay of the land, weather, soil acidity and having a design that makes sense to that place – it's important to take the time to do a land observation. The following is a basic design workshop which is a good way to create a collective vision of what you want to do in a few hours. It could be expanded to fill a whole day or a smaller group could expand on the plans. The designs could be displayed in a public place for comments and to get more people interested in the project.

Box 10.2 Design workshop for community garden

1. Introduce the idea and share ideas about all the benefits of a
    community garden.
2. Ideastorm the things people would like to see in the garden
    such as: vegetable patch, sensory garden, wildlife areas/pond,
    leisure/barbecue area, comfrey/nettle/wildflower patch, kids
    area, workshop area, seating/shaded area, shed/polytunnel/
    greenhouse/indoor area for bad weather, security hedges, raised
    beds so that elderly and disabled people can also participate in
    the garden.
3. A design tool frequently used by organic and permaculture
    gardeners known as OBREDIM helps to plan the design. It stands
    for Observations, Boundaries, Resources, Evaluation, Design,
    Implementation and Maintenance.

Observations. This stage is potentially endless as there
will always be things changing. In small groups fill in
an observation sheet looking for the following:

  * Access to the land
  * Plants already growing – this can help you work
    out loads of things about the land from soil quality and type,
    humidity, etc.
  * Any signs of wildlife present
  * Sun and prevailing wind direction
  * Water supply and possible water collection points
  * Shaded areas from buildings/trees
  * Slopes and rough/smooth/rocky areas
  * Wet/dry/swampy areas
  * Signs of contamination/nearby roads.

Pool the information that everyone has found onto a large, rough
map of the site.

Boundaries. What boundaries are there to the land, both physically
(waterways, hedges, trees, existing structures, slopes) and more
generally (financial, opinions of people nearby, etc.).

▷

Resources. What resources exist? Plants, water points, soil, 'rubbish' that could be recycled as plant pots, materials for mulching, people.

Evaluation. Ask the group to get back into their smaller groups and draw their own site maps with what they have observed.

Design. Start to design the garden on site maps. Present the different design ideas and discuss them. Think about what is feasible and how these ideas can be implemented.

The next stages take months and years to develop, but hopefully the workshop will have stimulated interest and imagination and you will see the people at the next community garden work day.

Implementation. Once you have volunteers you can make a list of what needs to be done: building a compost bin, preparing the land, general maintenance work, sourcing seeds, etc. You will more than likely need seed trays, tools, string, thick material, old carpet and wood so start gathering useful materials from wherever you can – car boot sales, the skip and the dump are all good places to find things. Developing a feasible work plan that takes into account everybody's other commitments is important.

Maintenance. In the next section we will look at ideas for getting the garden going and growing.

how does your garden grow? ideas for your garden

Mulching

Covering the land you want to grow with mulch is an excellent way to start off a community garden – it's not technical but it has immediate benefits and gets people thinking about soil, water, light and bugs. Many materials can be reused as mulch such as straw, chipped bark, stable sweepings, lawn clippings, sawdust, newspapers, cardboard, leaf mould, seaweed, pine needles, nutshells, clothing, stones, old carpet and roofing underfelt. Mulching has many functions:

★ Prevents weeds from growing as there is no light
★ Keeps the soil at a more constant temperature

☆ Prevents water from evaporating and keeps the soil moist (i.e. less watering)

☆ Prevents vital minerals from being zapped by strong sun

☆ Improves soil (biodegradable mulches) when it breaks down

☆ Defines areas that are to be planted and pathways.

Start mulching by laying down a layer of wet newspapers and then placing material on top of it to a depth of 6 inches. Make sure that all the existing vegetation is covered. When the mulch is in place, you are ready to sow by tearing a hole in the bottom layer of newspaper, adding a handful of ripe compost, and planting the seed or seedling in the small mound. When the mulch breaks down it can be dug into the soil and helps improve soil quality. Clover, alfalfa or black plastic are also commonly used as mulch.

Crop rotation

This is the practice of planting crops in a different plot each year. There are a number of reasons to do this: it prevents diseases building up in the soil, controls weeds

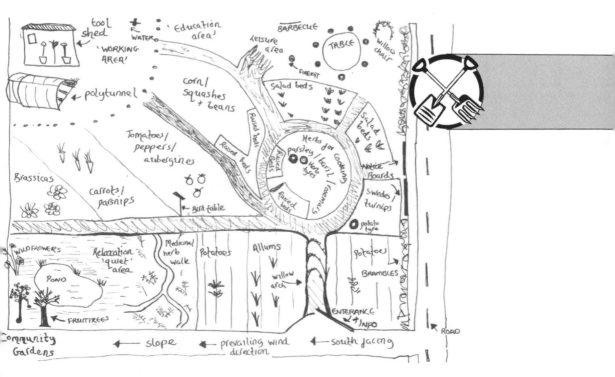

Figure 10.1   Design for a community garden

Source: Kim Bryan.

by regularly changing their growing conditions and prevents the soil becoming exhausted.

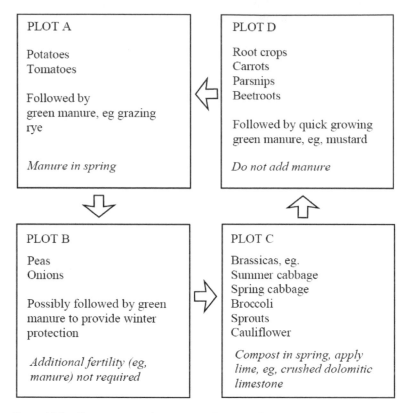

Figure 10.2   Crop rotation. An example of a typical four-year system

Source: Spiral Seeds.

Tiered growing

Often we think about gardening taking place on the ground but by being a bit creative it's possible to grow things all over the place. Especially in smaller gardens tiered plants look great as well as maximising space. One well known example is 'the three sisters' (corn, squashes and beans) which all grow really well together. Plant the corn and beans first and let them establish themselves, and after around three weeks plant out the squash. The corn grows up to about 2 metres, the beans wind themselves around the maize as they need to climb and the squashes grow on the ground. It's a perfect relationship.

Potato tyres

Take an old car tyre, fill it with a nice healthy soil and plant a couple of potatoes that have seeded. When you see green leaves appearing, place another car tyre on top of the first one and half fill that with soil. As soon as green leaves appear again, fill the rest of the tyre. When green leaves appear, place another car tyre on top of the first two, and so on. All being well, when the white flowers die back on the plant (normally 5 car tyres high) you should have a bumper crop of potatoes. The tyres can also be painted to make interesting garden sculptures.

Archways

Climbers can be trained along willow rods of equal lengths formed into an archway. Cucumbers, beans and tomatoes also like growing up walls, sheds, fences and trees. Train the plant up the willow and during the summer you will have created a shady, cool archway.

Companion planting

This is the practice of planting species in close proximity to each other so that they benefit each other. It works for several reasons; some plants are more attractive to pests than others and 'trap' insects, stopping them from eating other more needed or useful crops. Plants such as peas, beans and clover keep nitrogen in the soil which other plants need to grow. Other plants exude chemicals that suppress or repel pests and protect neighbouring plants. Tall-growing, sun-loving plants may share space with lower-growing, shade-tolerant species, resulting in higher total yields from the land. They can also provide a windbreak for more vulnerable species. Companion planting works because it encourages diversity. The more you mix crops and varieties the less chance you have of losing all your crop. For example, the cabbage family is well companioned by aromatic herbs, celery, beets, onion family, chamomile, spinach and chard, while tomatoes grow well when planted near nasturtium, marigold, asparagus, carrot, parsley and cucumber.

Making simple garden compost

Composting kitchen waste used to be commonplace and is a really important way to cut emissions of methane from landfill sites. Even if people don't have time to work in the garden, many would be happy to see their vegetable peel turned into a lovely rich humus.

You need a site that's at least 1 m by 1 m and a container (see below). Start by spreading a layer that is several inches thick of coarse, dry brown stuff, such as leaves, twigs or old newspapers, and top that with several inches of green stuff (grass or plant cuttings). Add a thin layer of soil. Add a layer of brown stuff. Keep layering the compost heap in the same way and every couple of weeks use a garden fork or shovel to turn the pile. If you turn the pile every couple of weeks and keep it moist, you will begin to see earthworms throughout the pile and the centre of the pile will turn into black, crumbly, sweet smelling soil. When you have enough finished compost in the pile to use in your garden, shovel out the finished compost and start your next pile with any material that hadn't fully decomposed in the previous one

Building a compost bin

Many councils will provide free bins but it's also easy enough to build your own. Wire mesh compost bins are versatile, inexpensive and easy to construct. A circular wire mesh bin may be made from poultry wire, hardware cloth or heavy wire mesh.

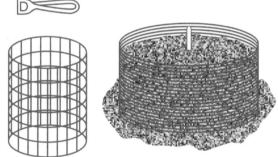

Figure 10.3   Circular wire mesh compost bin

Source: Adapted from Composting for All.

Four wooden pallets can also be hinged or wired together to construct a compost bin. The bin should be constructed with at least one removable side so that materials can be turned easily.

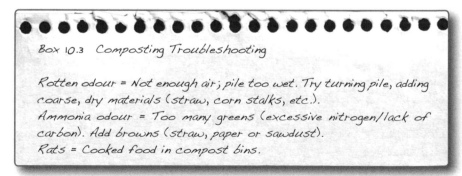

Box 10.3   *Composting Troubleshooting*

Rotten odour = Not enough air; pile too wet. Try turning pile, adding coarse, dry materials (straw, corn stalks, etc.).
Ammonia odour = Too many greens (excessive nitrogen/lack of carbon). Add browns (straw, paper or sawdust).
Rats = Cooked food in compost bins.

Source: Adapted from Composting for All website by Nicky Scott.

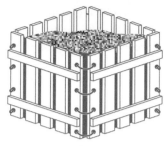

Figure 10.4　Wooden pallet compost bin

Source: Adapted from Composting for All.

the problem is the solution

'The problem is the solution' is a permaculture principle which states that by anticipating problems we can learn how to deal with them – we can turn problems to our advantage by being intent on creating solutions. Here we look at some common problems and solutions.

Poor quality soil

Soil type and quality are arguably the biggest determinants of your crops success. If the land has been steadily depleted over the years by the application of pesticides and fertilisers or is too stony, sandy or acidic there are a number of things that can be done to help the soil regenerate. Green manures (nutritional rich plants), such as

Box 10.4　Case study: community composting schemes

Micro-processing of your own waste is about taking direct responsibility for your actions. Yet many people do not have the space for compost bins. Community composting schemes are the perfect solution and provide compost for local parks and gardens. A successful group in London (East London Community Recycling Project, ELCRP) promoted their project as a way to reduce the smell of rubbish in estate stairwells and have had a very high take-up rate – 80 per cent in some cases. As well as encouraging residents to keep their food waste and advising them how to store it, the scheme offers regular doorstep collections. (www.elcrp-recycling.com)

alfalfa and clover, grow quickly providing a 'living' mulch that helps to recondition soil depleted by putting in valuable nutrients. Applying seaweed, which is rich in all trace elements and minerals, also helps. Earthworms improve soil fertility through aeration, drainage and by incorporating organic matter. Adding lots of compost will increase their numbers.

Figure 10.5   Earthworm

Source: Graham Burnett.

Contamination

Find out as much as you can about what the land has been used for in the past. If the land is contaminated it can be dangerous to eat foods grown there. Identify the source of contamination and assess whether it is liable to reoccur. Sources are most likely air, upstream water, imported products and landfill. Solutions are dependent on the type and severity of the contamination. Trees with high water uptake, such as poplars and willows, absorb toxic water through their root systems and break down toxic compounds inside tree tissue. Mustard plants and corn are able to absorb heavy metals and pollutants from the soil. Raised beds using large plastic/porcelain/wooden tubs and filled with soil from elsewhere can be built in order to grow food. Although they require more watering, raised beds are also good for kids, the elderly and people with mobility problems as they require less bending and are less easily trampled.

A weed is just a plant in the wrong place!

Nettles probably make most people think of stings, but the nettle is a great example of a common plant considered by many to be a weed which to those in the know is a multifunctional miracle plant! Nettles attract butterflies and moths (which eat aphids, which eat cabbages and beans), have a high vitamin C content, are a compost activator, can be used to make rope fibre, have many health

Figure 10.6   Stinging nettle (*Urtica dioica*)

Source: Graham Burnett.

benefits as a herbal medicine/tea, make an effective liquid fertiliser and can be used to make soup.

Fat hen is entirely edible and can be used as a spinach substitute or eaten raw in salads. The leaves are applied as a wash or poultice to insect bites, sunstroke, rheumatic joints and swollen feet, a green dye is obtained from the young shoots while the crushed fresh roots are a mild soap substitute.

Figure 10.7   Fat hen
(*Chenopodium alba*)

Source: Graham Burnett.

Pests

You will inevitably be sharing the plot with a variety of bugs and grubs. But even those that are pests have their part to play, such as providing a source of food for beneficial insects; so aim to manage them rather than wipe them out. Many conventional pesticides kill everything, good and bad. Provide habitats to attract insects by planting small flowering plants. Keep down dust and provide water as it attracts insects. An old baby bath can be a replacement pond and hedges provide a barrier to prevent dust and pollution settling on plants. Learning to recognise who is who will help. For example, discourage millipedes, which eat bulbs, potatoes and plant roots, but encourage centipedes, which are fast moving predators that live on small slugs and other soil pests.

Aphids are tiny insects that form colonies that suck sap. The larvae in particular will debilitate plants especially broad beans and brassicas but they are eaten by ladybirds.

Figure 10.8   Aphid

Source: Graham Burnett.

Figure 10.9   Ladybird

Source: Graham Burnett.

Any gardener will tell you how much they dislike slugs as they chomp their way through a garden eating seedlings and plants. Here are some tried and tested methods for slug management in the garden:

★ Slugs love certain plants and totally ignore others. Ones they love include any type of squash or courgette and mint. They hate onions and garlic (maybe it's a breath thing!). Work out which ones they particularly go for and concentrate slug defences in those areas. Larger slugs prefer dead vegetation and are less of a problem.

★ Slugs can hide under the mulch. Go slug hunting on wet days as this is the type of weather they love the most and will come out from under the mulch.

★ The sweet intoxicating smell of flat beer entices slugs and they come in their droves, drowning drunk must be better than drowning sober.

★ A plastic bottle with the top cut off can be used to protect plants from slugs.

★ Slugs cannot crawl over salt and ash so try sprinkling around crops.

★ Ground beetles are also predators of slugs, mites and other pests. Create a habitat for them by leaving pieces of wood for them to shelter under.

Figure 10.10   Slug

Source: Graham Burnett.

Figure 10.11   Beetle

Source: Graham Burnett.

Unwelcome guests of the human variety

Whether kids really are worse these days or it just seems that way, it's a sad fact that community gardens can be an easy target for bored kids. Many community gardens and allotments are protected by metal fences with spikes, which not only characterises a fearful and defensive society, but can also be seen as a challenge to bored young people looking for something to do. There's no easy answer to the carrot pulling youth, a much maligned and misunderstood part of any community. Remember, people that feel excluded from the garden won't respect it. Try making the garden an obvious community project through signs, lots of open invitations and inclusive design and implementation processes. Find space and offer to co-organise events that they may come to such as DJing or graffiti competitions. Invite the local school for a tour or work session and offer neighbours produce in return for a watchful eye. Red tomatoes falling from the vines are tempting trouble, so harvest all ripe fruit and vegetables on a daily basis. Fences serve to mark possession of a property as well as prevent entry. Nothing short of razor wire and land mines will keep out a really determined person, short picket fences or chicken wire will suffice to keep out dogs. Make fences attractive but thorny by planting fruit bushes such gooseberries or blackberries.

conclusions

Living gardens are places of learning, social contact and connection and are fun, practical and very necessary. As well as all of the benefits it brings to the environment and your community it can also benefit you – by being outdoors, digging, planting and taking time to observe patterns and changes in nature, seeing seasons change from one to the other. Everyone benefits from a connection with nature, but it's something that can frequently be put on the back-burner due to the stresses of everyday life. Urban areas have lots of energetic people who, when putting their minds and hearts together, can create beautiful, healthy, living environments. And bear in mind these golden rules: start small, build up gradually, take it easy and enjoy it!

Alice Cutler and Kim Bryan both campaign and facilitate educational projects around climate change and sustainability issues. Alice has been involved with several community garden projects and is a regular cook at the Cowley Club cafe in the Cowley Social Centre in Brighton. Kim is a trained permaculture designer and teacher with experience working in organic, community gardens and a number of different land projects including Escanda in Spain (see www.escanda.org). Many thanks to all the people that contributed to this chapter, particularly Ruth and Graham Burnett.

resources

Books

Andrews, Sophie (2001). *The Allotment Handbook – A Guide to Promoting and Protecting Your Site*. Bath: Eco-logic Books.

Barclay, Eliza (2003). *Cuba's Security in Fresh Produce*. Oakland, CA: Food First Institute for Food and Development Policy, www.foodfirst.org/node/1208

Burnett, Graham (2000). *Permaculture: A Beginners Guide*. East Meon, Hampshire: Permanent Publications.

Burnett, Graham (2000). *Getting Started on your Allotment.* Spiral Seeds. The Westcliff Land Cultivation Society. East Meon, Hampshire: Permanent Publications.

Butler, T. and K. McHenry (2000). *FOOD NOT BOMBS. How to Feed the Hungry and Build Community.* Tuscon: Sharp Press.

Carter, Warren. *Seedy Business: Tales from an Allotment Shed,* www.seedybusiness.org

Cherfas, Jeremy, Michel Fanton and Jude Fanton (1996). *Seed-Savers' Handbook.* Thuder Bay, Ontario: Grover Books.

Diamond, J. (1997). *Guns, Germs, and Steel: The Fates of Human Societies.* New York: W.W. Norton and Co.

Ellis, Barbara W. and Fern Marshall Bradley (1996). *The Organic Gardeners Handbook of Natural Insect and Disease Control.* New York and Pennsylvania: Rodale Press.

Ferguson, Sarah (1999). *A Brief History of Community Gardening in NYC, Avant Gardening.* New York: Autonomedia.

Fern, Ken (1997). *Plants for a Future: Edible and Useful Plants for a Healthier World.* East Meon, Hampshire: Permanent Publications. Also see online database www.pfaf.org

Funes, Fernando, Luis García, Martin Bourque, Nilda Pérez and Peter Rosset (2005). *Sustainable Agriculture and Resistance: Transforming Food Production in Cuba.* Oakland, CA: Food First Books and ACTAF.

Guerra, Micheal (2000). *The Edible Container Garden: Fresh Food from Tiny Spaces.* London: Gaia.

Hart, Robert (1996). *Forest Gardening.* Totnes, Devon: Green Books.

Henry Doubleday Research Association (1997). *Encyclopedia of Organic Gardening.* London: Dorling Kindersley Publications.

Hessayon, D.G. (1997). *Vegetable and Herb Expert.* London: Sterling.

Holt-Gimenez, Eric (2006). *Campesino a Campesino: Voices from Latin America's Farmer to Farmer Movement for Sustainable Agriculture.* Oakland, CA: Food First Books.

Jones, J.A. (1999). 'The Environmental Impacts of Distributing Consumer Goods: A Case Study on Dessert Apples'. PhD thesis (unpublished). Centre for Environmental Strategy, University of Surrey, Guildford, UK.

Larkcom, Joy (2001). *The Organic Salad Garden.* London: Francis Lincoln.

Lawson, Laura J. (2005). *City Bountiful. A Century of Community Gardening in America.* Berkeley, CA: University of California Press.

Manning, Richard (2004a). 'The Oil We Eat'. *Harpers Magazine* 808, www.harpers.org/ TheOilWeEat.html

Manning, Richard (2004b). *Against the Grain: How Agriculture has Hijacked Civilization.* New York: North Point Press.

Payne, D., D. Fryman and D. Boekelheide (2001). *Cultivating Community: Principles and Practices for Community Gardening as a Community-building Tool.* Philadelphia, CA: American Community Gardening Association.

Rosset, Peter and Medea Benjamin (1994). *Cuba's Experiment with Organic Agriculture.* West Sussex: Ocean Press.

Shiva, Vandana (1997). *Biopiracy: The Plunder of Nature and Knowledge.* Cambridge, Mass: South End Press.

Stickland, Sue (1998). *Heritage Vegetables.* London: Gaia.

Stickland, Sue (2001). *Back Garden Seed Saving: Keeping Our Vegetable Heritage Alive.* Bath: Eco-logic Books/Worldly Goods.

Ward, Colin and David Crouch (1998). *Allotments, Landscape and Culture.* Nottingham: Five Leaves Books.

Whitfield, Patrick (2004). *Earth Care Manual.* East Meon, Hampshire: Permanent Publications.

Reports

Church, Norman. *Why Our Food is so Dependent on Oil,* www.energybulletin.net/5045.html

DETR (2000). Department of the Environment Transport and the Regions. *Focus on Ports.* London: HMSO.

Ethical Trading Initiative (2003). *Report on the ETI Biennial Conference 2003. Key Challenges in Ethical Trade.* London: ETI.

Food Poverty Project (2002). Eds A. Watson et al. *Hunger from the Inside: The Experience of Food Poverty in the UK.* London: Sustain. (See http://worldcat.org/wcpa/oclc/59333724)

IUCC (1993). Information Unit on Climate Change Fact Sheet 271, 1 May 1993.

National Centre for Health Statistics (1998). *10 Million Americans of All Ages Do Not Get Enough to Eat,* www.cdc.gov/nchs/pressroom/98facts/foodinsu.htm

Shrybman, Steven (2000). *Trade, Agriculture and Climate Change.* Institute of Agriculture and Trade Policy.

Sustain. *Changing Diets, Changing Minds: How Food Affects Mental Health and Behaviour,* www.sustainweb.org

Sustain/Elm Farm Research Centre (2001). *Eating Oil – Food in a Changing Climate,* December 2001, http://worldcat.org/wcpa/oclc/59435091

Vetterlein, John. *How to Set Up a Food Co-op,* www.upstart.coop/page36.html

Websites

American Community Gardening Association www.communitygarden.org (A superb website with lots of hands-on advice and publications.)

Can Madeu Social Centre www.canmasdeu.net

Community Composting Network www.communitycompost.org (Supports projects and has a library of books, display materials, presentation materials and videos.)

Community Supported Agriculture Scheme www.hillandhollowfarm.com/csa.html

Composting for All www.savvygardener.com/Features/composting.html

East London Community Recycling Project www.elcrp-recycling.com

Food Agricultural Organisation www.fao.org

Food Not Bombs www.fnbnews.org

Guerilla Gardeners www.guerillagardening.org

Hartcliffe Health and Environment Action Group www.hheag.org.uk

Hedgehogs www.uksafari.com (Adopt a hedgehog.)

Institute for Agriculture and Trade Policy www.iatp.org

International Movement for Food Sovereignty www.viacampensina.org

Landless Peasant Movement: Brazil www.mstbrazil.org

Moulsecoomb Forest Garden and Wildlife Project www.seedybusiness.org

National Allotments Association UK www.nslag.org.uk (For support and local information.)

National Food Alliance Food Poverty Project www.sustainweb.org/poverty_index.asp

Permaculture Association UK www.permaculture.org.uk (For a comprehensive list of permaculture courses, online resources and links.)

Primalseeds www.primalseeds.org.uk (Network to protect biodiversity and create food security.)

Slow Food Movement www.slowmovement.com

Spiralseed www.spiralseed.co.uk

Sustain www.sustainweb.org (Alliance for better food and farming.)

Union of Co-operative Enterprises www.cooperatives-uk.coop

Wwoof (Willing Workers on Organic Farms) www.wwoof.org (Database of organic farms; welcomes volunteers.)

11 why we need cultural activism

Jennifer Verson

cultural activism: direct action and full spectrum resistance

By the left, quick march...

In July 2005 it seemed almost normal for me to be with 200 other people all dressed in crazy camouflage adorned with pink and green fluff, clown white on our faces and colanders on our heads. On our way to meet the Make Poverty History marchers in Edinburgh during the 2005 G8 summit to invite them to join us in taking direct action, we were trying hard to march in step with our feather dusters strung over our shoulders. OK, the Clandestine Insurgent Rebel Clown Army is an absurd army but it is also a serious response to the criminalisation of protesters and dissent. So I was not prepared for the angry comments of a very serious acquaintance of mine, dressed in black: 'This isn't funny. We are at war and we need to be able to fight. ... You are encouraging people to think that this is a joke, and the state is very, very serious.'

I'm worried. If we are at war, then it doesn't seem that we are winning. Is my friend suggesting that our failures as a movement are partly my fault? Are all of the clowns, drummers, pink and silver ballerinas and puppeteers, the cultural activists responsible? Do we encourage people to deal with the deadliness of the state war machine in an unrealistic way? This is what runs through my mind constantly these days.

This chapter is about cultural activism and I explain this isn't just about making things pretty, fluffy or fun. Cultural activists are taking direct action against war, ecological destruction, injustice and capitalism, but they are also constantly asking how we can act directly against their social and psychological effects. Just as military empires have defined full spectrum domination, we have embraced the idea of full spectrum resistance. After all, who can really know what it is that really inspires an individual to care, or to turn away, to give up or to rise up?

Figure 11.1   Rebel clown at the G8 summit, Scotland 2005

Source: Guy Smallman.

To me, cultural activism is where art, activism, performance and politics meet, mingle and interact. It builds bridges between these forms but also exists as the bridge itself, stuck permanently between two places. What links activism and art is the shared desire to create the reality that you see in your mind's eye and believe in your capacity to build that world with your own hands.

Who am I to talk?

I see myself as part of a community which defines itself very much through its practice. Starting from what I know through personal experience, I will endeavour to expose some trends in cultural activism as well as some of its rich history which has been important to me. In doing this, I am pulling together theories that have influenced me from fields that vary widely – quantum physics, organisational psychology, postmodernism and permaculture – and which have been useful in helping me to understand why this cacophony of colour, sound and dissent is vital in changing the world.

so what is cultural activism

Cultural activism is difficult to define. The definition of activism in the *Oxford English Dictionary* is 'the use of vigorous campaigning to bring about political or social change'. But everybody from anthropologists to artists has been arguing for at least a hundred years on definitions of culture that range from the poetic to the straightforward. The noted anthropologist Clifford Geertz in *The Interpretation of Cultures* (1973, 5) says 'man is an animal suspended in webs of significance he himself has spun, I take culture to be those webs'. While UNESCO in its Universal Declaration on Cultural Diversity is far more direct and broadly encompassing: 'culture should be regarded as the set of distinctive spiritual, material, intellectual and emotional features of society or a social group, and ... it encompasses, in addition to art and literature, lifestyles, ways of living together, value systems, traditions and beliefs' (UNESCO 2001, 12). Aime Cesair, a Martiniquan writer, speaking to the World Congress of Black Writers and Artists in Paris, is the most direct: 'Culture is everything' (Petras and Petras 1995, 54). In this sense, cultural activism could use everything as a potential resource. For me, cultural activism is campaigning and direct action that seeks to take back control of how our webs of meaning, value systems, beliefs, art and literature, everything, are created and disseminated. It is an important way to question the dominant ways of seeing things and present alternative views of the world.

What ties together the myriad of forms that I will discuss in this chapter is that they take place not only in physical space but also in cultural or idea space. The following exercise (see Box 11.1) is adapted from the Smartmeme Collective's useful worksheet which helps ground grassroots activists in an understanding that power structures can be successfully challenged in a variety of ways (see www.smartmeme. com/downloads/InterventionsWorksheet.pdf).

Box 11.1   *The points of intervention*

This exercise is intended to help grassroots activists identify points of intervention in both physical and cultural or idea space where they can take action in order to make change. Points of intervention are a place in a system, be it a physical system (chain of production, political decision making) or a conceptual system (ideology, cultural assumption, etc.), where action can be taken to ▷

effectively interrupt the system. The points which are considered in the exercise are:

Point of Production
Factory, crop lands. The realm of strikes, picket lines, crop-sits, etc.

Point of Destruction
Resource extraction such as clear cuts mines, etc. Point of toxic discharge, etc. Realm of road blockades, tree-sits, etc.

Point of Consumption
Chain stores, supermarkets. Places where customers can be reached. The realm of consumer boycotts and market campaigns.

Point of Decision
Corporate HQ. Slumlord's office. Location of targeted decision maker.

Point of Potential
Future scenarios, actualising alternatives, transforming an empty lot into a garden, Reclaim the Streets, etc.

Point of Assumption
Challenging underlying beliefs or control mythologies, such as environment must be sacrificed for jobs. Also hijacking spectacles and using popular culture.

The old resistance of barricades, marches or armed guerrilla groups intervened frequently at the points of production, destruction or decision. The forms that we look at in this chapter are interventions at points of potential, assumption and consumption. This is a more savvy resistance which uses our media saturated society to develop new forms of actions that are ever shifting, mutating and always one step ahead of those who want to co-opt and restrain us. To understand what cultural activism means in more detail, below are some key aspects we need to consider.

Insurrectionary imagination

If you woke up one morning and decided to do something that you had never done before you would probably look for guidance. If you bake a chocolate cake you would probably get a recipe. Emboldened by your cake success you decide to build a straw bale house, so you go to the library and get a book with instructions, diagrams and pictures. Baking a cake and building a house are infinitely easier than creating a sane and just world, so how can we be expected to do it without instructions? It is no wonder that people love political manifestos that prescribe exactly how to do it. But what would happen if you sat down and visualised the world that you wanted to live in? Were you even ever taught how to visualise? Have you practised closing your eyes and seeing a picture in your mind? What if you not only could see that picture clearly but also truly believed that you could create it in your waking life?

An insurrectionary imagination is at the heart of cultural activism. It is a sense of possibility that is not limited by copying a pattern or following a design that somebody else created or by what Augusto Boal (2002) calls the 'cop in the head'. We all have that voice, the one that tells us our ideas are stupid, they won't work out, they are too difficult or are bound to fail. Cultural activism relies on killing the cop in your head and expressly tries to develop this insurrectionary imagination to create performances and actions. This living practice addresses complicated questions about how we build the world that we want to live in. Insurrectionary imaginations evoke a type of activism that is rooted in the blueprints and patterns of political movements of the past but is driven by its hunger for new processes of art and protest.

Dialogue and interactivity

Giving people long sermons on the need for them to get involved in change can often be patronising and disempowering. Traditional campaigning tends to involve attempting to attract people to a cause by bombarding them with facts and fiery speeches. Cultural activism tends to move away from one-sided monologues, speeches and propaganda into porous forms that use dialogue and interaction. Is it such a huge stretch of the imagination to believe that people can speak for themselves and ultimately have something to say about the world they want to live in?

Community, concrete action and campaigning

Even though people may have passionate opinions and beliefs about the world they want to live in there are some very real obstacles that stand in the way of people organising and acting for local and global change. Psychologically, we have all experienced feelings of despondency, apathy and helplessness, while physically many of us are isolated from other people, as well from ideas, resources and information that can help us. Cultural activism, in all its varied forms tries to address these issues through:

✭ Community: Access to time and space free from consumerism, coercion and capitalism for conviviality, alliance building and information sharing.

✭ Concrete Action: Doing something, anything – whether it is repairing a billboard or hosting an open stage performance – breaks the social conditioning of complacency, despondency and apathy.

✭ Campaigning: Access to information that is not censored by the corporate limited media.

We can all change the world, can't we?
The title of this book 'Do It Yourself: A Handbook for Changing our World' reflects a seismic shift in thought that is pervasive in many of the above forms – focusing on

Box 11.2   *The five Cs of cultural resistance*

Artist activist John Jordan, a veteran of cultural resistance strategies, illuminated some of these patterns through the five Cs of cultural resistance:

The Courage to disobey. The courage to realise that the only thing to fear is fear itself. The courage to follow our coeur – our heart – that lies at the root of the word 'courage'.

The Creativity that dares to dissolve the boundary between dream and reality. The creativity that believes in the unlimited power of our imaginations. The creativity that is the key that opens the door to our unbounded freedom.

The Craft that takes time to sculpt and form ideas and materials. The craft that values precision and complexity, and that knows that the magic and beauty of everything lies in the detail.

The Commitment to follow an intuition and idea to its end. The commitment that never gives up or goes home. The commitment to hold on to our truths in spite of ridicule, scepticism and attack.

The Cheek that always remembers, with a wry smile, to pour gallons of pleasure and play into every one of our acts of creative resistance.

'our world' rather than 'the world' and the power of the individual to create change rather than relying on the 'masses'. Many cultural activists approach social change with an ecological perspective, viewing both movements and power structures as holistic systems where all the parts are interconnected.

In searching for political metaphors other than that of 'the masses' many cultural activists have embraced metaphors and paradigms that validate their beliefs that one person can change the world. The variety of cultural actions can act like:

★ a lever that moves a huge boulder with very little force
★ a spanner in the works of a corporate machine
★ a butterfly flapping its wings that creates a hurricane.

Fritjof Capra in *The Web of Life* (1997, 132) explains the latter:

Chaotic systems are characterized by extreme sensitivity to initial conditions. Minute changes in the system's initial state will lead over time to large scale consequences. In chaos theory this is known as the butterfly effect because of the half-joking assertion that a butterfly stirring the air today in Beijing can cause a storm in New York next month.

This kind of metaphor of change was summed up beautifully by Patrick Reinsborough of the Smartmeme Collective: 'what has mass impact doesn't necessarily need masses of people' (interview with the author, July 2006).

roots and shoots

How has cultural activism worked out in practice over the years and in different places? There is a huge array of important influences to be looked at from political theatre to visual art, and social movements. In this section I link these historical roots with more contemporary shoots which have drawn on these historical examples in some way. I look at examples in the three areas of art, theatre and carnival that have either had strong influences on me or are part of major trends in current activist practice.

1. Fine art and public art

Roots: Muralist movement

It is doubtful that any revolution in any country ever had such a talented, perceptive group of artists to record its struggle for freedom. (Smith 1968, 281)

In the 1920s, in post-revolutionary Mexico, artists such as Diego Rivera painted politically charged murals depicting scenes of the revolution and indigenous life before the conquistadors. The Mexican muralists both told the history of the nation and created a national mythology, painting on public buildings as an attempt to unite and educate the people. Jose Vasconcelos, the Minister of Education after the revolution, set up a network of rural art schools. His vision of Mexican identity as a 'cosmic race' synthesising native, European, African and Asian cultures was brought to the cultural or idea space directly in the muralists' work. In 'Birth of a Nation', for example, Rufino Tamayo paints powerful images of Mexican identity, depicting a woman giving birth to a baby, half red, half white, as she is trampled under the hooves of the conquistador.

Murals reclaim public space and use it to tell and celebrate histories and struggles. The aesthetics of the Mexican Muralists have inspired artists around the Americas, Europe and Russia. The belief that the visual space in the public realm belongs to the people has fuelled directly or indirectly a wide spectrum of practices.

Shoots: Subvertising   In the urban areas of the 'overdeveloped' countries, one of the most visible forms of cultural activism is anti-capitalist actions against corporate culture. Subvertising, often included under the umbrella term of 'culture jamming', refers to making spoofs or parodies of corporate and political advertisements in order to make a statement. Media vary from crudely altered billboards and lamp post stickers to T-shirts and TV ads. According to *AdBusters*, a Canadian magazine and a leading proponent of counterculture:

> a 'subvert' mimics the look and feel of the targeted ad, promoting the classic 'double-take' as viewers suddenly realize they have been duped. Subverts create cognitive dissonance. It cuts through the hype and glitz of our mediated reality and, momentarily, reveals a deeper truth within.

Geert Lovink, media theorist, net critic and activist, comments:

> In my view culture jamming is useless fun. That's exactly why you should do it. Commit senseless acts of beauty. But don't think they are effective, or subversive, for that matter. The real purpose of corporation cannot be revealed by media activism. That can only be done by years long, painstakingly slow, investigative journalism. Brand damage has never been proven enough. What we need is research and thinking, brainstorming, and then action. (www.networkcultures.org/geert/speed-interview-conducted-by-andre-mesquita-brazil)

Figure 11.2   Subvertising, Manchester, UK

Source: Indymedia UK.

While some question the usefulness of this type of media activism it is crucially changing both idea space and public space from a corporate or party political monologue to a dialogue where people are speaking for themselves. Pat Tinsley, the operation superintendent of Eller Media, an outdoor advertising company in Oakland, California, concedes his antagonists are disciplined. His employees have spent long hours reconstructing the altered signs. 'It's a high cost to us, but they are creative and use their imagination', he says. 'They are very professional. They use rigs, they use the right glue, they operate at night in areas where they know there will be few cars going by. When it happens. we say, "Our friends have struck again".'

It is not only an accessible form – billboards advertising military tattoos, four-by-fours or election propaganda are irresistible canvasses that can be 'repaired' with a few swooshes of a spray can. But also, the humour (at least in Oakland, California) builds bridges instead of antagonism.

2. Political theatre

Roots: Theatre of the oppressed   In the early 1970s Brazilian director Augusto Boal developed a pioneering body of work which struck at the historic relationship between the stage and spectator to create an arena where oppressed people could become actors in their own liberation. In *The Rainbow of the Desire* (1999), Boal tells a story about his first plays – idealistic, leftist works that explained the necessity of a social revolution to workers and peasants. One day, after calling people to arms in his performance, a peasant took the message literally and suggested Boal and his troupe join him in getting weapons and killing the landlord. Boal had a hard time explaining that they were simple actors and not fighters. At that moment, he realised that he was using his theatre to ask others to do what he was not willing to do himself.

The practice that developed from this realisation is called forum theatre where a group of people from an oppressed group create a play about a specific situation. The scenario is played out with the predicted negative outcome; for example, a chauvinist man mistreating a woman or a factory owner mistreating an employee. The play is then performed for the community and the audience are 'spect-actors'. They are encouraged to interrupt the action and play out the possible changes on stage, and they then rehearse the change that they want to see.

Shoots: Forum theatre   'What defines Forum Theater as Theater of the Oppressed [is] its intention to transform the spectator into the protagonist of the theatrical action and by this transformation to try to change society rather than just interpreting it' (Boal 2002, 253). The shoots of Boal's early work are in two main camps. One is the proliferation of forum theatre that is progressively removed from the performance arena into more social contexts. Forum theatre is used everywhere – from workshops in primary schools addressing bullying to community events addressing domestic violence. Groups like the Cardboard Citizens in the UK address issues of homelessness, while in Port Townsend, in the USA, the Mandala Center uses forum theatre in anti-racism work for white people.

The other main strand can be seen in character-based action performances. These forms of cultural activism involve the protesters adopting 'characters' for specific direct actions. Most frequently the 'actors' are drawn from groups that are campaigning, rather than from those with an acting or performance background. In some of the most visible groups irony is the tactic of choice: CRAP, Capitalism Represents Acceptable Practice, does action or performances where a group of (mostly) 'non-actors' play 'capitalists' carrying signs that say 'Fuck the Third World' and 'bombs not bread'. These forms differ from both carnival and clowning (see below) where you strip away the masks and artifices of society, as they draw on the historic theatrical tradition of 'character' where you adopt the persona of somebody other than yourself.

3. Carnival

Roots: The carnival tradition   'It is necessary to work out in our soul a firm belief in the need and possibility of a complete exit from the present order of this life'. So Mikhail Bakhtin quotes Dubrolybou in *Rabelais and His World* (1984, 274) to explain how carnival was necessary in the Renaissance movement to overcome the weight of the medieval cultural view of the world as unchanging.

In medieval Europe the carnival square was a place of freedom, collective ridicule of officialdom, celebration, feasting and breaking the normal social restrictions of both hierarchy and decorum. From the seventeenth to the twentieth century there were 'literally thousands of acts of legislation introduced which attempted to eliminate carnival and popular festivity from European life' (Stallybrass and White 1993). In South America and the Caribbean, carnival, though nominally introduced by the Catholic colonists, had taken root and flourished and married itself to African traditions of parading, costume, music and mask.

Carnival enabled people to be able to envision and believe in a different type of society and, according to Bakhtin, individual thinking and scholarly writing were not enough: 'popular culture alone could offer this support' (Bakhtin 1984, 275). People who believe in carnival say it is about rehearsing what it is like to be free, a time when power is inverted and the world is turned upside down. Sceptics about carnival argue that it is catharsis, a time for the oppressed people to blow off steam so that they are willing to accept their lot in life for the rest of the year. Whichever your view, throughout history carnival has been a time for inverting the social order. There is something carnivalesque in many of history's unpredictable moments of rebellion, from the French revolution to the civil rights, suffragette and anti-slavery movement 'where the village fool dresses as the king and the king waits on the pauper, where men and women wear each others' clothing and perform each others' roles. This inversion exposes the power structures and illuminates the processes of maintaining hierarchies; seen from a new angle, the foundations of authority are shaken up and flipped around' (Notes from Nowhere Collective 2004, 174–5).

Shoots: Tactical frivolity   One common feature of marches and mobilisations has become carnival forms. The iconic colours of pink and silver infuse protest with playful costumes, puppets, singing, chanting and dancing. Some will say that pink and silver is making protest 'fun' but to me it's about tapping into ancient forms of collective celebration, which are about inclusion and joy, something we often lack in individualised Western cultures. Music has always been a way of expressing dissent and strength when there is little other way to do it. From the chain gangs, slaves and mine workers to rastas and rappers, from punk rockers to peace ballads, trade union brass bands to the New Song movement, music and dance has been key to building social movements. Samba bands draw on a long tradition of carnival, such as the UK action samba group, Rhythms of Resistance, who use the slogan 'Only Music Can Save Us Now'. Bands bring vibrancy to protest marches comprised of diverse people, whose outrage cannot be encapsulated in a unifying chant or slogan. Samba bands have been used to move large groups of people in street protests, to block roads and occupy buildings, and for noise demonstrations and solidarity actions outside police stations.

This brief overview shows a wide range of examples. I want to now look at a recent phenomenon of cultural activism that I have been involved in, that of rebel clowning.

Figure 11.3   The pink and silver block

Source: Guy Smallman.

the emperor has no clothes: the emergence of the rebel clown

In the lead-up to the anti-capitalist and global justice protests against the G8 summit, Scotland, in July 2005, a group of artists and activists toured the UK to apply creativity to radical social and ecological change. The Laboratory of Insurrectionary Imagination tour was a combination of live art and a cultural resistance training camp, which combined street interventions, a nomadic information centre and two-day intensive rebel clown training.

A group with more than 40 combined years of experience in clowning, collective theatre, improvisation and direct action delivered workshops designed to quickly move people towards the physical understanding of a few key concepts:

- ✭ spontaneity (to kill the 'cop in the head')
- ✭ complicity (to practise radical democracy and play)
- ✭ releasing the clown.

Many of the great clown teachers, like Jacques Lecoq and Phillip Gaullier, teach that everybody has at least one clown inside of them and that clown training is about 'releasing the clown', removing the barriers that society has put in place to keep us from our authentic self.

Box 11.3   Basic rebel clown improvisation training

In basic rebel clown training we did improvisation training that I call Yes, No, But, And.

Step 1: A invites B to do something, B says No.
For example:

A: Would you like to go for a walk?
B: No. (A keeps inviting, B keeps rejecting.)

Step 2: A invites B to do something, B says Yes, then makes an excuse.
For example:

A: Would you like to go for a walk?
B: Yes, but I don't have any shoes.
A: How about a barefoot walk on the beach?
B: Yes, but I don't have any sunscreen.
A: Then how about we go after sunset?
B: Yes, but I will be hungry then (etc.).

Step 3: A invites B to do something, B not only accepts the offer but makes an additional suggestion.

A: Would you like to for a walk?
B: Yes, and we can go to the beach!

▷

A: Yes, and when we get to the beach we can have a picnic!

B: Yes, and we can invite other people on the beach to share our picnic!

A: Yes, and the people sharing our picnic can tell us stories about their lives, etc.

The point of the exercise is to show that by saying 'No' and 'Yes, but' we stifle not only our own imagination and initiative, but that of our friends and colleagues.

Rebel clowning, an emergent form that combines the ancient art of clowning with the practice of non-violent civil disobedience, resulted from a wide variety of performers and activists saying 'Yes and ...'. Rebel clowning is partially a tactical weapon against the sheer stupidity of capitalism and war, and partially a tool to free the self from the tremendous damage that capitalism has done to our bodies and our minds.

How does it work?

Fishing is a good example of the possibilities of rebel clowning to create both personal and political change. On the second day of clown training groups begin to learn some basic rebel clown manoeuvres. Drawing on the tradition of lazzi in *commedia de l'arte*, performances are improvised but the performers have some stock gags that they can repeat. Instead of stock gags, the clown army has manoeuvres. Fishing simply entails the entire group moving through space, tightly packed together doing synchronised movement. As the group turns, the person who finds themself in front becomes responsible for initiating a new movement and sound. The leader organically rotates. Practising this physical form of rotating leadership erodes assumptions about leadership and hierarchy that are deeply ingrained in our very bodies.

One of the most magical moments I have ever seen was in the Scotland G8 protests when a group of clowns was being penned in by riot police on horses, and the clowns were fishing down the street in perfect synchronicity, clippity clop, clippity clop, clippity clop, doing a collective impression of a horse. As one of the major functions of mounted police is to intimidate protesters, the act of laughing at the horses was significant in redistributing power and agency in the situation. Unlike the story of Boal above, where the performers shy away when the farmers want to take immediate action, those of us travelling the UK, recruiting and training rebel clowns were actually calling for people to take part in impending direct actions that we ourselves would be participating in. This is an exciting development across the art and activism field, where the art in itself is inseparable from the direct action.

But how is fishing going to save the world?   Having trained in exercises such as fishing, by July 2005 over 300 clowns were prepared for making decisions without a leader, as well as using physical instead of verbal communication once the protests happened. What did this lead to? Clowns appeared to be everywhere, the clowns could not be contained, the clowns as an entity had become stronger than its parts. On the other hand, a group of police went home and told the story about how, during the protests, they played a game of giants, wizards and goblins with a group of clowns! Who knows where that will lead? When the convergence camp in Stirling during the G8 summit was surrounded by police, a group of clowns faced police officers who had been trained to deal with them. 'I am not going to fall for your tricks', a police person growled at me. Whatever we were doing, I thought, it must have been working. Clowning is dangerous because it subverts the protocols of war and policing; where 'good' protesters are supposed to obey authority and 'bad' protesters are supposed to resist with violence. It deconstructs the opposition between fluffy and violent protest. Our desire to be free is not funny, it is war. But at the same time, I still think it's kind of funny that the state is threatened by clowns.

Consuming solutions: 'Those fucking clowns!'
Ideas are not harmless. Activists need to be careful and question the whole nature of ideas being viral, because freedom must involve thinking for yourself. This is why real engagement and creativity is so vital to the ideas of cultural activism, and should never be let go even if it seems easier just to copy someone else's idea. We have seen this with most forms of cultural activism – as people are hungry for answers and solutions they are quick to take up the next big trend in protest movements. While rebel clowning and clown training can be a way of healing the body and the mind from the social damage of capitalism, if it becomes a virus people will copy what they have seen either at protests, at the circus or on television. They will cut out the development of the insurrectionary imagination, and it is the development of this imagination that opens up limitless worlds of activism beyond costumes and stock gags.

There is also a long history of co-opting countercultural fashions and ideas into the mainstream by marketing professionals. Community art can be socially conservative, questioning some values, while replicating others. The idea of community art has been used as a pawn in urban 'renewal' to make 'dangerous' neighbourhoods safe for 'development'. All too often the search for space has made artists and activists complicit in a brutal gentrification process that benefits only property developers.

new metaphors for cultural action

> Emancipate yourself from mental slavery; None but ourselves can free our mind.
> (Bob Marley, 'Redemption Song')

We need new metaphors, new ideas and new images to understand ourselves and how we are going to create our everyday revolutions. And, like everything else, we are going to have to create these new metaphors ourselves. Back in the old days, with the old thinking of us versus them, people may have asked 'How is one garden going to change the world?' In the new times with new metaphors of ecological thinking (we are all in the same boat and it is sinking, so we had better do SOMETHING!) we can understand that one garden is the world changed. All of the practical suggestions in this book, all of the gardens, health clinics and protests, are not possible unless each of us frees our minds, believes in our own potential and our own power. Capitalism does its best to make us feel helpless – 'Got to pay the rent and keep a roof over my head' – and convince us that we need to dominate others (both as individuals and as nations) to ensure our own freedom. Cultural activism is necessary to give us all the mental and physical tools we need to free our bodies and our minds. We don't just rehearse the revolution, we practise it everyday.

Jennifer Verson is a freelance performance activist from the USA who trains people in the art of politically subversive theatrical performances and who worked extensively with the Clandestine Insurgent Rebel Clown Army touring the UK in 2005.

12 how to prank, play and subvert the system

The Vacuum Cleaner

introducing: the barratt* homes of creativity

It didn't take long. By the time Gap's broken windows had been replaced, advertising types had appropriated the images of anti-WTO protesters smashing glass in the Battle of Seattle, 1999. Diesel swiftly launched its protest chic range, whilst Boxfresh splashed Zapatista leader Subcomandante Insurgente Marcos on T-shirts for £25 a pop. The revolution will be commercialised. Che Guevara T-shirt anyone?

Images and stories are the way most people will encounter this movement of resistance and creation. One placard that got plenty of media coverage at the biggest 2003 anti-war march in London was 'Make Tea Not War', a placard, it turned out, that was made by an advertising agency. The question is: whose propaganda wins? The black-clad unwashed or the Diesel model dressed up to look like they give a shit? Whose side of the story is being told? What effect do charity celebrities such as Bob Geldof and Bono have on radical agendas, and on the way our stories, desires, rage and alternatives are presented?

This movement can and does create powerful images and stories that are vital tools for radical change. When we think about creativity we don't mean paint and canvas, bright costumes or painting your face. And by images we don't just mean photographs. What we feel is important is what our stories do and how they are framed. Is the story moving? Is the experience aesthetically pleasing? This may seem a trivial question as global melt approaches ever faster. But, these days the battles happens more in the mental environment of advertising space and public relations trickery than in the streets. On the street activists can score well, but on the other front we are unprepared and losing. Yet crafted new forms of resistance are engaging

---

* Barratt is a British construction company famous for building millions of houses that look the same all over the islands.

people and interrupting the system with unpredictable tactics that can't always be shut down quickly. This approach to an action or event is key in the battle of the story and image, and to the ways in which our actions and beliefs are understood by the people we encounter.

We started to make activist art after hearing stories of a particular, exciting performance. A group known as Fanclub went out shopping. They bought things and then returned them repeatedly, over and over, monopolising the queue, holding shops to ransom with their own logic. Intrigued, we began looking around and discovered other similar groups and actions, such as the Surveillance Camera Players who play in front of CCTV and Whirl-Mart Ritual Resistance, who block the aisles of supermarkets, going round and round. The stories of these groups, seeing their rebellious images, which were as seductive as any advert, revealed to us a world of possibilities for cross-pollination of politics and art forms.

For us, the most difficult thing was finding the confidence to try something new. Our friends told us to 'Just Do It', follow Nike, it doesn't matter if it doesn't work the first time. So, on Buy Nothing Day 2002, we organised a party in the Virgin Megastore in the shopping mecca of Camden Town, London, under the name CCCP (Camden Corporate Crackdown Party). As we had feared it turned out to be a crap action. Only three people turned up, not much of a party. But we had tried and this spurred us on to organise the next Whirl-Mart Ritual Resistance actions, which we started in the UK in 2002. We haven't stopped since.

We share four stories to inspire and motivate. The first story we experienced first-hand, the second we were involved in creating, and the last two we watched from afar with great interest and excitement. We don't believe we can tell you 'how to' subvert the system. We are working on the assumption that the stories we tell could trigger something in your imagination.

story 1: shop lifting with the church of stop shopping

The Reverend Billy and the Church of Stop Shopping are a theatre troupe-cum-direct action affinity group who have focused their actions mainly on Starbucks and Disney, companies with terrible environmental and labour records. Their actions are best described as part space hijacking and part church service. They regularly take over a Starbucks cafe or Disney store with hymns, sermons and exorcisms. We have always been big fans of their 'Church'. What has inspired us when we hear of their latest adventures (aside from the original idea of having a church against consumption, a 30-strong stop shopping gospel choir and Billy having a restraining order from every Starbucks in the USA) is that they keep what they do fresh and unexpected, constantly pushing in new directions. One example was on Boxing Day, 2005, when they hijacked the Christmas parade at Disney Land, Florida. Whilst innovative, the second great thing is their ability to not stop. Most likely we would have burned out by now or changed our angle. But, resembling a radical steamroller, the Church of Stop Shopping is persistent. They aren't perfect but they are radical both in their politics and their form.

What we'd like to share with you isn't Disney Land hijacks or Church services to thousands. It is something a little smaller. It was after the G8 summit in Scotland in 2005. The Church were in town. We talked about how the G8 had been, about Geldof and Bono hijacking the agenda, and the 7 July bombings in London before we got around to the topic of visiting a few Starbucks the next day. As we have mentioned, Billy is now banned from every Starbucks in the USA after allegedly 'sexually assaulting a till register' (he was actually exorcising it). They suggest various performance options before mentioning an action called *Shop Lifting*, at which our ears pricked up. Turns out to be shop lifting, the act of lifting the shop, rather than shoplifting, the act of liberating products. As they explained the plan, we felt that excited tingle and contacted everyone we could think of. We needed at least ten people to carry it off, and we were sceptical.

However, on the morning of the action there was a good crowd: clowns, environmentalists, squatters, academics, experienced space hijackers and some people who had just found out and come along. We started off by sharing our experiences of Starbucks, whether as activist, consumer or neither. It was a very simple exercise but quickly forged a sense of the group, what experiences and expectations people had, and how far people were willing to push things.

Starbucks is designed like an Ikea display. The coffee isn't the only exploitative product inside; everything from the door handles to the tables and chairs are equally unsustainable. Tracing the origins of products is hard, if not impossible. Yet each object has a story, a point of origin from which we can trace a clear form; a chair was once a tree, a napkin also, a plastic bag was once oil deep in a well. Our action focused on these distant origins, tracing these histories and making them seen, heard and spoken. We did this at the journey's current end, a Starbucks on New Oxford Street, one of the biggest outlets in London.

At intervals the group enters Starbucks and selects an object they would like to trace. Placing ourselves in the middle of a group of tables, we drag a free chair into the middle of the space and begin. Others are standing in front of the counter, some by the bins, one in front of the door blocking the entrance. There are as many of us as there are customers. We begin quietly, taking our time, following every point of the object's journey; from the van it arrived in, the warehouse-lorry-boat-lorry-warehouse-lorry, and so on, lifting it every time. The chair that we are lifting is at waist height and we are only on the boat, the chair is nearly above our heads, and we still have a way to go before becoming a tree. Already a few customers have walked out, a lot have walked past, some are ignoring us, others are watching with great interest. Each person, and each history, is getting louder and louder until the music can't drown us out and this space isn't a third place anymore; it is made up of trees, drops of oil in a well and coffee beans. Reverend Billy storms in, blond quiff almost two steps ahead: 'Brothers and Sisters of Starbucks...', he begins his now well-rehearsed sermon about the evils of transnational coffee hell, explaining how on this day we have imagined a new beginning for the objects in this Starbucks. He goes on... this is a chance for the sinners (shoppers) to move on from the shiny veneer of bullshit, the third place, just off third way. For this moment, the space is ours, everyone has stopped what they are doing, time itself has slowed down. The whole of New Oxford Street seems to know that something different is happening, that the boredom of fashion and fad is momentarily over and that a group of people has found a new religion, the Church of Stop Shopping. The non-shoppers put back the objects where they got them and calmly leave the store. Looking back over our shoulder, the store seems deserted.

We move on to the next Starbucks, conveniently less than 15 metres away, and repeat the action. This time the police are called and arrive very promptly, only to stand and watch, paralysed by a lack of understanding or a fear of being made to look more stupid than we do already. Twenty people holding chairs, cups and coffee above their heads, screaming 'I am now a tree again!' We disappear into the crowds of people who have gathered outside before regrouping a few blocks away and dispersing to share stories for the rest of the afternoon.

We've seen Billy do much pushier things. However, seen as part of a body of work that the Church has undertaken for years, the enormous effect that this group has had becomes clear, and all from having an idea and running with it. By keeping the work fresh and accessible Reverend Billy and the Church of Stop Shopping are responsible for a popular alternative that has spread to groups of people who may have never encountered this type of politics in this way before. It's also possible that it has had an impact on the store itself.

story 2: the church of the immaculate consumption – praying to products

The Church of Stop Shopping tries to raise awareness about the stupidity of mindless consumerism. An offshoot, the Church of the Immaculate Consumption encourages believers to worship the product rather than buy it. As most people in late capitalist societies understand, consumerism is no longer about getting what you need to survive but about reaching godliness, personal fulfilment, through the purchase of lifestyle. Most people in these societies also understand the iconography of the Christian religion, believers or not. The Church of the Immaculate Consumption in large cathedrals of consumption (otherwise known as shopping centres) started as small groups quietly getting down on their knees and asking for salvation and fulfilment from the product. Two years later, the Church was holding some of the biggest stores in the UK to ransom with their own logic.

In early 2003, a group gathered around some frozen chickens and prayed, 'Asda, thank you for these lovely, lovely, fresh, white chickens at cheap, cheap prices.' About ten people saw it first-hand. It was pretty low-grade on the civil disobedience scale but it spread as a story and a short video.

Later we repeated the action in the flagship House of Fraser store, Glasgow, and Brent Cross Shopping Centre, London. However, it didn't really take off as a mass effective action until the following year during the 2004 Autonomous Spaces sessions at the European Social Forum in London, as part of the Laboratory of Insurrectionary Imagination's (Lab of II) programme of events. 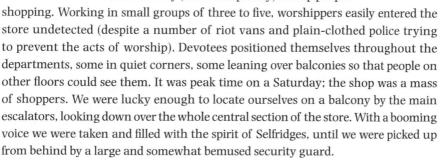 By this point we'd organised a few more events and were more confident. The target was the Selfridges store, the jewel in Oxford Street's crown. With around 50 people taking part the action was transformed into an effective way (albeit temporary) to stop people shopping. Working in small groups of three to five, worshippers easily entered the store undetected (despite a number of riot vans and plain-clothed police trying to prevent the acts of worship). Devotees positioned themselves throughout the departments, some in quiet corners, some leaning over balconies so that people on other floors could see them. It was peak time on a Saturday; the shop was a mass of shoppers. We were lucky enough to locate ourselves on a balcony by the main escalators, looking down over the whole central section of the store. With a booming voice we were taken and filled with the spirit of Selfridges, until we were picked up from behind by a large and somewhat bemused security guard.

Figure 12.1   Praise be, to therapy

Source: The Reverend Billy.

Groups prayed in their own way, some loudly, some not. After being removed by security staff, groups were able to re-enter through different entrances and continue their prayers. The store was only left with one option, to stop letting everyone in as they couldn't possibly determine who was a shopper and who was part of the congregation. Rumours vary but some say it was an hour before the doors reopened.

However, this wasn't enough for the Church members. The next level was to repeat the success of the Selfridges action weekly. In the nine cities we visited on the Lab of II tour before the 2005 G8 summit in Scotland, we always headed down to the local multinational cathedral to worship. In this way we could hit the chain in a sustained campaign. Each action was unique, some more successful than others. But each congregation was able not only to hold these spaces to ransom but also to share skills, support each other, and find a point where they were comfortable performing and push themselves from there.

story 3: nikeground – guerrilla marketing or collective hallucination?

Picture This: Rethinking Space. Having the chance to redesign the city where you live, waking up in the morning and changing the names of streets and squares, to imagine bizarre monuments and then the next day see them become real; it's the dream of every citizen, it's Nike Ground! This revolutionary project is transforming and updating your urban space. Nike is introducing its legendary brand into squares, streets, parks and boulevards: Nikesquare, Nikestreet, Piazzanike, Plazanike or Nikestrasse will appear in major capitals over the coming years. (www.nikeground.com)

Picture this: The square and piazzas of major cities, public spaces famous for their monuments, mass demonstrations and tourist hot spots, often with historic names – Trafalgar Square after the battle of Trafalgar, St Anna's Square after St Anna, Union Square after the famous union marches – these places are steeped in history and collective memory, not just for local people but also globally through images and footage. However, with increasing privatisation of public space should place names not begin to reflect the importance of the corporation? Should we not consider a programme of renaming? Starbucks Square, for example, or McDonald's Passage? The Nike Corporation, which has always been at the cutting edge of marketing, has already begun this vandalism of our physical and mental space. Vienna, Austria, was the scene of a battle between the local Viennese and Nike's public relations machine.

Nike, without any consultation of the public, announced that the historic square, Karlsplatz, right in the heart of historic Vienna, was to be renamed Nikeplatz. Nike's famous 'Swoosh' logo would become a giant monument to the corporation and this historic occasion. To mark the event, Nike would open an exhibition centre in Karlsplatz promoting the virtues of having a big name brand as the focus of the city centre. A launch event invited locals to come and learn about Nike's plans, with freebies and Nike beauties extolling the brand's virtues. A new trainer was even launched to commemorate the rebranding of the square. The press were notified and, of course, flocked to cover the story.

This would have been an activist nightmare had it been real. It was actually a brilliant prank pulled by the Italian duo known as 0100101110101101. The group had secured permission from the Viennese authorities to install their exhibition by the roadside in Karlsplatz, and the towering Nike 'Info Box' looked 100 per cent authentic. Combined with a website, www.nikeground.com, and a few slick press

Figure 12.2   The proposed Nike monument in Karlsplatz, Vienna

Source: www.0100101110101101.org.

releases, the group had secured an outpouring of hate, shock, disdain and maybe even love, towards Nike for suggesting such an idea.

After a few months the press exposed the Nikeground campaign as a stunt. But, as any public relations person will testify, the damage had already been done. In the minds of Vienna's consumers Nike had attained a negative status, the brand has been damaged and maybe even lost some of its devotees. Nike's PR people pulled out all the stops to deny, what was for all intents and purposes, their own propaganda. Resorting to legal action against the pranksters was their only option for rebuke but they eventually dropped the case. One can only wonder that they feared ending up in a PR disaster to rival McLibel. In that now infamous legal battle between two UK activists and McDonald's, the fast food giant exposed its own darkest secrets in an attempt to silence protesters.

The high level of press coverage was what made the prank a success, yet all the pranksters had done was taken the identity and logic of Nike and then pushed it to its next step. Through this 0100101110101101 had revealed the true, hideous nature of corporate globalisation.

story 4: dow chemicals correction

On 2 December 1984 a chemical leak from a Union Carbide factory released 27 tonnes of methyl isocyanate gas throughout the city of Bhopal, India. Since that night, it has killed 20,000 people and a further 120,000 people have suffered serious illness, blindness and reproduction problems as a direct result. Despite a sustained campaign by the people of Bhopal to demand a clean up and calls by the Indian government to extradite Warren Anderson, then Chief Executive, on multiple

Figure 12.3
The Yes Men
live on TV

Source: The Yes Men.

homicide charges, nothing has been done by the company to care for those who were affected. Although Union Carbide, now Dow Chemicals, had let safety at the plant go out the window, to this day they have done nothing to clean up or pay any compensation and US law enforcement has done nothing to turn Warren Anderson over to the Indian authorities.

In November of 2004, almost 20 years after the accident, a representative of Dow Chemicals finally got around to apologising on BBC World News. Jude Finisterra said sorry for doing nothing for so long and announced that Dow would pay $12 billion compensation to the victims and remediation of the site and waters of Bhopal, the exact amount that Dow had paid for buying Union Carbide. They also stated that they would push for the extradition of Warren Anderson to India, after he had fled to the US following his arrest 20 years ago. Within minutes, the news of Dow's policy shift became a major news story. On the German stock exchange Dow Chemicals' value plummeted, losing the company $2 billion.

Acceptable risk

Six months earlier another Dow representative, Erastus Hamm, had spoken at a banking conference in London. During his presentation he spoke about a new Dow industry standard known as 'Acceptable Risk'. This is a statistical equation to determine how many deaths are acceptable in the pursuit of large profits. It was designed to help the corporation deal with the skeletons in their closet. Mr Hamm reminded delegates that they could not all assume to be as lucky as IBM, who made a killing from its computer systems sold to the Nazis for counting the Jewish, gay, disabled and other 'expendable' people into the death camps. The presentation involved the unveiling of a golden skeleton to help describe the idea of financial profit from loss of life. Numerous bankers signed up for a licence to use this new method.

As you have probably guessed by now, both Acceptable Risk and the apology turned out to be the work of the Yes Men who had impersonated their representatives in the guise of Mr Finisterra and Mr Hamm. Unsurprisingly, Dow Chemicals were not best pleased when it was revealed that a group of pranksters had not only been able to speak on their behalf, but dramatically rocked their share price, and gone on to paint them as a company that doesn't see any problems in profiting from killing people. The Yes Men shot to stardom when they released a film about their pranks, speaking on behalf of the World Trade Organisation.

The AGM

A year after launching Acceptable Risk and it is time for the Dow Chemicals Annual General Meeting (AGM). What better opportunity than the AGM of shareholders to wrap up a series of stunts? As it turns out the Yes Men aren't the only activist group planning to attend and security is tight; all journalists have been banned, everyone is searched on the way in and any recording equipment confiscated. The security guards tell the Yes Men that they like their film, which they have been shown to emphasise the potential threat. But because they have one share in the company the guards have to let them in. Determined to have the last laugh they decide to give it a go; after the usual speeches and a few questions Andy (a Yes Man) is introduced as Jude Finisterra, the same name he gave for the BBC interview:

> Hello Bill, shareholders. We made an incredible $1.35 billion this quarter. That's really terrific. But you know, for most of us, that'll just mean a new set of golf clubs. I for one would forego my golf clubs this year to do something useful instead – like finally cleaning up the Bhopal plant site, or funding the new clinic there. Bill, will you use Dow's first-quarter profits to finally clean up Bhopal?

The question is quickly fielded away, but Mike (another Yes Man) has the next question in the queue:

Great profits. Now I wanna see you use them to go after some of the creeps who are tarnishing Dow's good name! I'm looking around and most of the questions are from people who don't like Dow. Let's do something about that. We need to get aggressive! Of course you can't exactly broadside a bunch of nuns with a twenty-gun shoot, and you can't just kick a disabled kid in the head – but at least you could take care of hooligans, like that guy who went on the TV news to announce that Dow was liquidating Union Carbide. That made a serious stock bounce, and I for one was freaked out! I'll bet a lot of you were! So, Mr Stavropoulos, are you going about that criminal? And if not, why not?

And the response? 'If you can tell us who that guy is. Next question?'

Over the period of a year the Yes Men had managed to present Dow Chemicals in a completely new light across the world's media, costing the company probably millions of dollars, and putting Bhopal back on the media agenda in the USA.

conclusion: we suck

Think about times when someone has tried to explain or promote an idea, ideology or possibility to you. Imagine a conversation where you are a Buddhist, and a black bloc anarchist tries to defend property damage, or that you are an eco-terrorist and your uncle is an oil baron. It is quite possible that you were mutually unsympathetic and at the end of the day, remained unconvinced. The moments we remember listening really hard and being sympathetic to what is being said are times of funny stories, sad moments or seductive possibilities. What could be less appealing than being lectured about what is wrong with your lifestyle? How a message is presented is integral to its effectiveness. This is not to say that pranks and creative resistance are the only tools. Sometimes a march from A to B can have a much bigger effect; other times it's as dull as voting. Sometimes a leaflet can cause major ripples, sometimes no one will read it. Sometimes a lock-on is the only thing that will work. But either way if we only rely on what is the done thing or what someone else has tried and tested before, then the power of that tool is reduced. It is easy to categorise, dismiss or ignore it. After all, fresh bread tastes better than stale.

Box 12.1   *Five tips for effective interventions*

1. Bear in mind the timing and location of your prank – the number of people can be the difference between a hit and a miss.
2. Do research and share knowledge of the target.
3. Get to know each other, your strengths, weaknesses and limits.
4. Take corporate messages to their extreme conclusions – when seen through a magnifying glass most people will find them unappealing.
5. Document your action – in high resolution. Subversion is contagious.

The Vacuum Cleaner are an artist activist collective of one based in Glasgow, Scotland. They make live and digital actions, interventions and pranks, and can be found at www. thevacuumcleaner.co.uk.

resources

Books

Auslander, Philip (1994). *Presence and Resistance: Postmodernism and Cultural Politics in Contemporary American Performance*. Ann Arbor, MI: University of Michigan Press.

Bakhtin, Mikhail (1984). *Rabelais and His World*. Trans. Helene Iswolsky. Bloomington, IN: Indiana University Press.

Bernays, Edward (1928). *Propaganda*. New York: Liveright.

Boal, Augusto (2002). *Games for Actors and Non Actors*. London: Routledge.

Capra, Fritjof (1997). *The Web of Life*. London: Flamingo.

Duncombe, Stephen (ed.) (2002). *Cultural Resistance: A Reader*. London:Verso.

Fo, Dario (1998). *The Tricks of the Trade, Independence*. Kentucky: Routledge.

Geertz, Clifford (1973). *The Interpretation of Culture*. New York: Basic Books.

Jasper, James M. (1998). *The Art of Moral Protest: Culture, Biography, and Creativity in Social Movements*. Chicago, IL: University of Chicago Press.

Kershaw, Baz (1992). *The Radical in Performance: Between Brecht and Baudrillard*. London: Routledge.

McGrath, John (1989). *A Good Night Out*. London: Methuen Drama.

Notes from Nowhere Collective (2004). *Carnival: Resistance is the Secret of Joy, We Are Everywhere*. London: Verso.

Petras, Kathryn and Ross Petras (eds) (1995). *The Whole World Book of Quotations: Wisdom from Women and Men Around the Globe through the Centuries*. Reading, MA: Addison-Wesley.

Ruby, K. (2000). *Wise Fool Basics: A Handbook of Core Techniques. Wise Fool Puppet Interventions*. Sante Fe: AK Press.

Schechter, Joel (1985). *Durov's Pig: Clowns, Politics, and Theatre*. New York: Theatre Communications Group.

Smith, Bradley (1968). *Mexico, A History in Art*. New York: Harper and Row.

Stallybrass, Pete and Allon White (1993). 'Bourgeois Hysteria and the Carnivalesque'. In *The Cultural Studies Reader*. Ed. Simon During. London: Routledge.

Talen, Bill (2003). *What Should I Do if Rev Billy is In my Store?* New York: New Press.

The Yes Men (2004). *The True Story of the End of the WTO*. New York: The Disinformation Company.

UNESCO (2001). Universal Declaration on Cultural Diversity. Adopted by the 31st session of the UNESCO General Conference, Paris, 2 November 2001, http://unesdoc.unesco.org/images/001271/127160m.pdf

Stories 1–4

Story 1. You can find the Church at www.revbilly.com. Bill Talen has also written a book *What Should I Do if Reverend Billy is In my Store* (which was the title of a memo sent to Starbucks in the New York area).

Story 2. For praying videos and more on these actions visit www.consume.org.uk

Story 3. 0100101110101101 have pulled numerous pranks, as we write they are currently releasing a film staring Ewan McGregor and Penelope Cruz, or are they? Find out more at www.0100101110101101.org

Story 4. Watch the BBC interview at www.theyesmen.org (The True Story of the End of the WTO – The Yes Men).

Websites

Magazines/journals

AdBusters www.adbusters.org

Community Arts Network www.communityarts.net/readingroom

Greenpepper www.greenpepper.org

www.journalofaestheticsandprotest.org

Collectives/individuals

Breathing Planet www.breathingplanet.net/whirl

Conglomo Media Network www.conglomco.org

Laboratory of Insurrectionary Imagination www.labofii.net

Mischief Makers www.mischiefmakers.org.uk

My Dads Strip Club www.mydadsstripclub.com

Northern Arts Tactical Offensive www.nato.uk.net

Not Bored www.notbored.org/the-scp.html

One-man subversive think-tank. www.dedomenici.co.uk

Rebel Clown Army www.clownarmy.org

Rhythms of Resistance www.rhythmsofresistance.co.uk

RTMark www.rtmark.com

Space Hijackers www.spacehijackers.org

Spanish brand of shoplifting www.yomango.netwww.yomango.net

Games. theatre, improvisation, making

Carnival: An Introduction to Bakhtin www.iep.utm.edu/b/bakhtin.htm

Puppets, 68 ways to Make Really Big Puppets www.gis.net/~puppetco/

www.improvresourcecenter.com

www.yesand.com (For sites on improvisational theatre and comedy.)

Memes

Memefest www.memefest.org

Smart Meme www.smartmeme.com

Theatre groups and forum theatre

Artists Network www.artistsnetwork.org

Bread and Puppet Theater www.theaterofmemory.com/art/bread/bread.html

Cardboard Citizenswww.cardboardcitizens.org.uk

Mandala Center for Change www.mandalaforchange.com

Toplab www.toplab.org

13 why we need autonomous spaces in the fight against capitalism

Paul Chatterton and Stuart Hodkinson

As we contemplate the vicious assault on even the limited democratic systems, rights and welfare we have historically won from capital, and as the spectre of irreversible climate change and 'peak oil' looms ever closer, the question of how we as ordinary people respond – and from what *bases* of resistance – becomes increasingly urgent. An important lesson from the history of popular struggles against the theft of common land, the wholesale destruction of communities or the imposition of authoritarian rule, is how they have often striven to create their own self-managed, autonomous spaces free of those very relations of domination and exploitation and as bases from which to regroup and politically organise. This idea can be found in land movements from the English Diggers to the Brazilian MST (Landless Workers' Movement), the European squatting movement, the occupied factories of Argentina and the Italian social centres. The term 'autonomous space' is of course a political one – there is no such thing as 'autonomy' under capitalism; there is no 'outside' the system. Autonomous spaces are instead places of creativity and experimentation where the colonising, dehumanising and exploitative logic of capitalism is actively resisted by people trying to live and relate to each other as equals. This chapter discusses the crucial role that fighting for such autonomous spaces can play today in both resisting global capitalism and helping us to develop viable alternatives to the private profit system. We begin by situating the idea of autonomous spaces within the historical struggle between Enclosures and Commons, or what we today might call the choice between privatisation and direct democracy.

resisting enclosure

The present crises of corporate power, the privatisation of our resources and the expansion of authoritarian social control are nothing new. The drive to 'enclose', 'dispossess' and 'enslave' has been a constant feature of human societies from the early empires of the Mayas, Aztecs, Romans and across Mesopotamia, to the violent birth pangs of global capitalism in the 1400s heralded by the start of the great land enclosures in England and the European slave harvesting from Africa, to the later city states and hinterlands controlled by wealthy burghs and the nobility respectively, and the more recent great westwards movement in the USA that disposed native Americans of their lands and created huge farming areas for European settlers.

Enclosures are deliberate and necessary mechanisms of domination, exploitation and power. The land enclosures in Britain from the fifteenth century onwards, led by the monastic orders, kings, warring lords, landowners and later the feudal manorial system, forced farming communities off their land and commons, which were then enclosed, privatised and capitalised as giant agricultural areas to enrich their new owners. By robbing people of their land this 'primitive accumulation' ended the communal control of the means of subsistence, separated people from these means, and created a population of workers with only their labour to sell as a means of survival. As capitalism spread across northern Europe, it was this landless working class that fed the needs of, first, agricultural capitalism and, then, the voracious demand of nineteenth-century industrial cities for cheap factory hands. The simultaneous development of European colonial systems in Latin America, Africa and Asia directly stripped indigenous peoples from their land, deprived them of their own basic resources such as timber, water, oil and minerals, imposed foreign rule and developed their economies as mere supply chains to enrich European societies.

Instead of being a one-off act of acquisition, however, the process of enclosure is a constant feature of capitalism in response to its internal contradictions. Since the global neoliberal turn from the mid 1970s onwards, international business and its state partners in government and the international financial institutions (IFIs) have orchestrated a dramatic and unprecedented enclosure of land and life in almost every corner of the globe, what is better known as 'neoliberal globalisation'. In the global South, the policies of privatisation, lower corporate taxes, public spending cuts, free trade and anti-trade unionism – often termed 'the Washington consensus' due to their championing by the US-dominated International Monetary Fund and World Bank – have merely replaced 'colonialism' and formal independence with 'neo-colonialism'. Direct military rule has given way to free trade areas and conditional debt relief as the main levers by which the rich North strips poor countries of their natural resources and wealth. Global corporations linked to ruling political elites, not nation states, have gained back control of industries at knock down prices, while deregulation has led to labour and environmental abuses as capital has been allowed to rip free. Seizing land for debt has pushed a massive new army of labour into the world economy, making mobile and migrant labour the dominant form, undermining collective organisation and place based struggles, depressing wages, and making workers vulnerable and precarious, and thus more compliant. The liberalisation of financial markets has enabled massive and almost instantaneous transfers of money, which have rapidly devalued and bankrupted economies. Thailand, Russia and Argentina literally lost control of their sovereignty, assets, resources and industries all in a matter of a few years (see Stiglitz 2002, Klein 2001). What we see is a more gloves-off type of global capitalism, based on more nineteenth-century style primitive forms of economic policy as corporations are able to dispossess and accumulate in brutal and ruthless ways (see Harvey 2005, Pilger 2002).

One of the most disturbing effects is growing volatility in terms of shelter and housing. The International Alliance of Inhabitants states that 15 per cent of the world's population is now threatened with eviction and the United Nations estimated that in 1996 1 billion people lived in inadequate housing (quoted in Corr 1999, 4). Furthermore, Mike Davis in *City of Slums* (2005) suggests that in Africa by 2015, there will be 332 million slum dwellers. Even for those with some money, price inflation and property speculation, as well as debt, has meant it is virtually impossible to buy your own house, let alone find land. This problem is acute in urban areas where land values have soared over the last decade due to the growth of large corporate-backed entertainment-retail-office-condo mega projects. Enclosure is now also spreading to the very building blocks of life as corporations claim legal ownership by patenting ancient knowledges, medicines and genetic plant codes. This is known as biopiracy (see Shiva 1997), the appropriation and monopolisation of traditional knowledges and biological resources leading to the loss of control of resources. In recent years, through advances in biotechnology, and international agreements on intellectual property such as TRIPs (trade-related aspects of intellectual property rights), the possibilities of such exploitation have multiplied.

These are not just developing world phenomena. Similar policies have ripped through the rich North. Downtown areas of big cities have become hubs for global capital and playgrounds for wealthy residents and business elites creating a doughnut effect between the hyper-rich core and the impoverished outer rim. At the same time, the state with its falling tax base is shaking off the management of public services like hospitals, schools, playing fields and parks, daycare centres and libraries. Its core business has now become attracting footloose investment, out-of-town tourists and business elites. As the welfare state crumbles, thousands have been left abandoned by a rolled-back state and a global economy that has moved elsewhere. What this adds up to is the destruction of the 'social commons'. Land and space is at a premium and goes to the highest bidder. As a result, community centres, local shops, post offices, public access woodland and open spaces, and working-class clubs all get sold off and recycled into shopping malls with private security guards, exclusive condos and riverside apartments or air-conditioned offices. Much of this is summed up by the term 'gentrification', where higher-order services and activities displace traditional, lower-order ones. These processes have been documented in excellent commentaries like Mike Davis' *City of Quartz* (1990) or Sharon Zukin's *The Culture of Cities* (1992).

In this perspective, we can see how enclosure and dispossession are widespread processes of commodification encompassing all of the commons – natural and human – from physical assets such as oil and public housing to the provision of welfare and health services; from open public spaces like city squares to ancient knowledges, medicines and the very genetic codes of life itself. In a world moment of potentially irreversible climate change and looming resource wars, the need to reclaim the commons from the rich and powerful, resist dispossession and enclosure, and create structures of solidarity and welfare, protection and shelter, has become nothing less than a question of human survival itself.

historical struggles for autonomous space

However, if enclosure, dispossession and enslavement have been constant throughout history, so too has the capacity and willingness of people to first resist and then seek ways to recollectivise their lives by managing and sharing land, work and resources together as equals. This democratic and egalitarian vision of 'autonomy' has its roots in the ancient tradition of the commons – the belief that the Earth and its resources belong to us all, and cannot be bought or sold in the marketplace, or claimed and partitioned by force for one group over another. It is thus a vision of self-management, non-hierarchy and mutual aid. The (re)claiming of space from private ownership by popular movements to recollectivise their lives and fight the commodification of land, labour and life has a long and rich history and offers inspiring examples.

The Diggers

In the wake of the land enclosures and destruction after the English civil war, the Diggers (or the True Levellers), led by the radical preacher Gerard Winstanley who believed that the Earth is a common treasury for all, established an agrarian community at St George's Hill in 1649 to feed and house the destitute. The experiment didn't last long and they were eventually evicted by the local lord, but it has had a lasting resonance as they advocated a radical restructuring of society where the poor would inherit the common wealth. Thousands of intentional agrarian communities have since been set up in reaction to the excesses of the Industrial Revolution and the prospects of a society without oil to work the land in an ecologically sustainable way. Echoing the sentiments of Winstanley, The Land is Ours group campaigns for the right to live sustainably on the land and promotes living on the land as well as land squats.

The European squatters

At its height in the 1980s, the European squatting scene was a mass movement, with some 3500 squatted houses in the Netherlands, 9000 in Berlin and 31,000 in London. Motives varied from the right to decent housing, challenging the corporate takeover of cities and the desire to set up self-managed social services in the gaps left by the crumbling welfare state. But the urban squatting movement also reflected a much wider rejection of life, work and politics under capitalism. While squatting is today in decline, in Europe some spaces have survived to become almost self-governing cities, such as Christiania in Copenhagen, dubbed 'Freetown' after it declared itself independent of the Danish state. Barcelona is also a great example of how a network of squats has formed to jointly advertise events and co-ordinate support.

The Italian Centri Sociali (social centres)

Some of today's autonomous self-managed spaces take their cue from Italy's Occupied Self-Managed Social Centres (CSOAs) that emerged during the extreme social unrest and economic crisis of the mid 1970s. They were founded by a non-parliamentary youth movement seeking to challenge a lack of housing, services and casualised labour, but rejecting both 'capitalist work' and the socialist parties who had abandoned working-class struggles for a share of state power. At the same, conventional political meeting spaces were being wiped out as workplaces, schools and universities closed. Enter the CSOAs. Taking over unused or condemned buildings, a network of autonomous cultural and political gathering spaces were created that have since become hives for experimentation with self management, independent cultural production of

Figure 13.1   The Square Social Centre, London, UK

Source: Paul Chatterton and Stuart Hodkinson.

music, 'zines, art and pirate micro TV and radio, publishing, and mutual aid welfare. Although the original social centres mainly died out in the repression of 1977, a new wave of 260 social centres emerged in the mid 1980s, the majority in Milan, Turin, Bologna and Rome. By 2001, this number had halved due to the dramatic shift towards neoliberal urban growth, while many of those that remained drifted into the mainstream as they became part of a normal night out for Italian youths and visiting tourists.

Travellers and ravers

The traveller community, long criminalised as deviants for their nomadic ways of life, know only too well the importance of who controls space. Clashes with the police over the right to live and party, and violent evictions have been frequent, especially in the UK in the mid 1990s. The rave and free party scene which swept the Western world at this time played a crucial role, squatting old warehouses and rural fields to self-manage huge music festivals and at the same time challenge the corporate control of music and lifestyles. However, laws like the 1995 Criminal Justice Act quickly legislated this growing movement out of existence, leading many to seek out freer spaces in southern Europe. Many others stayed and fought, such as the Exodus Collective in Luton who got involved in regeneration and turned around their local neighbourhood.

Anarchy in the UK: Punk, dole and the autonomy centres

In 1980s Britain, a heady mix of punk, anarchism and high unemployment led to an intense period of creativity and experimentation in the form of 'autonomy centres'. These clubs emerged in Britain's big cities as unemployed claimants tried to take back control of their lives against Thatcher's attack on working-class people, the betrayal by the Labour Party and official trade unions, and the rise of fascist activity. Inspired by the anger and energy of the punk scene through groups like Crass, and brought together through the Claimants' Unions, dozens of clubs opened up, such as the Autonomy Centre in Wapping and the 1 in 12 Club in Bradford. People got together to put on gigs, make 'zines, music, organise benefit advice, do anti-fascist politics, and get to know each other.

Indigenous land squats in the global South

Land squats, *favelas*, shanty towns, *villas miseries* – call them what you like – have become an everyday reality for indigenous peoples in the global South where access to land is a question of survival. They represent the growing insecurity of shelter and the normalisation of temporary and chronically poor slum developments for millions of people living on the edges of today's megacities. But some also represent living examples of how people can self manage their own communities even in extreme circumstances. In the 1960s there were over 1 million separate land squats, as revolutionary movements like the Sandinistas in Nicaragua made it their priority to redistribute and self-manage land. In Brazil, the MST (Landless Workers' Movement) has occupied and distributed land for 150,000 families since 1984 in a country with one of the worst distributions of land ownership in the world.

Protest camps and convergence sites

During the 1980s and 1990s, a broad direct action movement emerged in northern Europe around protest and peace camps determined to halt the expansion of roads, motorways, airports, military bases and nuclear weapons. The battles of Newbury and Twyford Down in Britain are famous for their dense resistance networks of tree houses, benders and mini camps, and their promotion of living, self-managed, ecological alternatives to the growth machine. The 2006 Camp for Climate Action outside

Drax Power station, the UK's single largest emitter of carbon dioxide, highlighted workable solutions to climate change. Peace camps, such as at Greenham Common and Menwith Hill, have confronted the war machine at its front door and spurred a whole generation of peace activists to peacefully resist nuclear missile programmes. More recently, 'No Border' camps have taken place along immigration borders under the banner 'No one is illegal'. During the summits of global and economic elites such as the G8, IMF and World Bank, large convergence centres and activist villages have been set up under names such as 'Inter-galactica' and 'HoriZone'.

Autonomous workers factories in Argentina

Argentina has been in crisis since 2001 when the country defaulted on its huge international loans, the government slashed the value of the peso, money and food ran out, and the rich took their money overseas. In response, many Argentinian workers reclaimed factories and businesses ranging from five-star hotels and pastry factories to metal works. Most organise using assemblies and flat structures to make decisions and have formed legally recognised workers' co-operatives or sought nation-alisation through government expropriation. Many of the reoccupied factories are co-ordinated through the National Movement of Recovered Factories, which has 3600 workers across 60 factories, and the National Federation of Workers' Co-operatives in Recovered Factories, which has 1447 workers across 14 factories. In a country where 60 per cent of people live below the poverty line, reclaiming a factory or business, running it autonomously and horizontally, and everyone getting fair wages is empowering and essential.

the role of autonomous spaces in today's resistance

What these selective episodes demonstrate is that when people work together they can challenge the status quo and start to take back control of their lives from the rich and powerful. We can also see, however, that the struggle for autonomous spaces and reclaiming the commons, whether land, housing, water, even the genetic codes of life, is never finished. The heyday of Italy's social centres has now passed, the recovered factories in Argentina face constant attempts by the bosses who abandoned them to seize them back, and squatting opportunities have become squeezed, especially in cities that have been rediscovered by rich elites as places to live, work and play. But efforts to set up self-managed autonomous spaces continue undimmed. Since the late 1990s, for example, the UK has seen the growth of a network of self-managed spaces and collectives (see Figure 13.2) broadly linked to the social centre idea, especially in downtown areas that have become more expensive, privatised and corporately

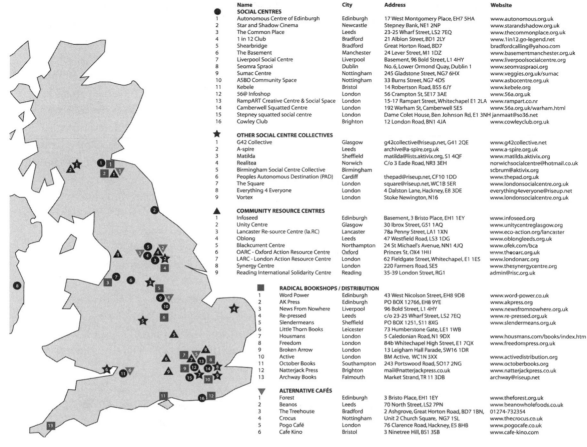

Name	City	Address	Website
SOCIAL CENTRES			
1 Autonomous Centre of Edinburgh	Edinburgh	17 West Montgomery Place, EH7 5HA	www.autonomous.org.uk
2 Star and Shadow Cinema	Newcastle	Stepney Bank, NE1 2NP	www.starandshadow.org.uk
3 The Common Place	Leeds	23-25 Wharf Street, LS2 7EQ	www.thecommonplace.org.uk
4 1 in 12 Club	Bradford	21 Albion Street, BD1 2LY	www.1in12.go-legend.net
5 Shearbridge	Bradford	Great Horton Road, BD7	bradfordcalling@yahoo.com
6 The Basement	Manchester	24 Lever Street, M1 1DZ	www.basementmanchester.org.uk
7 Liverpool Social Centre	Liverpool	Basement, 96 Bold Street, L1 4HY	www.liverpoolsocialcentre.org
8 Seomra Spraoi	Dublin	No. 6, Lower Ormond Quay, Dublin 1	www.seomraspraoi.org
9 Sumac Centre	Nottingham	245 Gladstone Street, NG7 6HX	www.veggies.org.uk/sumac
10 ASBO Community Space	Nottingham	33 Burns Street, NG7 4DS	www.asbocentre.org.uk
11 Kebele	Bristol	14 Robertson Road, BS5 6JY	www.kebele.org
12 56@ Infoshop	London	56 Crampton St, SE17 3AE	www.56a.org.uk
13 RampART Creative Centre & Social Space	London	15-17 Rampart Street, Whitechapel E1 2LA	www.rampart.co.nr
14 Camberwell Squatted Centre	London	192 Warham St, Camberwell SE5	www.56a.org.uk/warham.html
15 Stepney squatted social centre	London	Dame Colet House, Ben Johnson Rd, E1 3NH	janmaat@so36.net
16 Cowley Club	Brighton	12 London Road, BN1 4JA	www.cowleyclub.org.uk
OTHER SOCIAL CENTRE COLLECTIVES			
1 G42 Collective	Glasgow	g42collective@riseup.net, G41 2QE	www.g42collective.net
2 A-spire	Leeds	archive@a-spire.org.uk	www.a-spire.org.uk
3 Matilda	Sheffield	matilda@lists.aktivix.org, S1 4QF	www.matilda.aktivix.org
4 Realitea	Norwich	C/o 3 Eade Road, NR3 3EH	norwichsocialcentre@hotmail.co.uk
5 Birmingham Social Centre Collective	Birmingham		scbrum@aktivix.org
6 Peoples Autonomous Destination (PAD)	Cardiff	thepad@riseup.net, CF10 1DD	www.thepad.org.uk
7 The Square	London	square@riseup.net, WC1B 5ER	www.londonsocialcentre.org.uk
8 Everything 4 Everyone	London	4 Dalston Lane, Hackney, E8 3DE	everything4everyone@riseup.net
9 Vortex	London	Stoke Newington, N16	www.londonsocialcentre.org.uk
COMMUNITY RESOURCE CENTRES			
1 Infoseed	Edinburgh	Basement, 3 Bristo Place, EH1 1EY	www.infoseed.org
2 Unity Centre	Glasgow	30 Ibrox Street, G51 1AQ	www.unitycentreglasgow.org
3 Lancaster Re-source Centre (la.RC)	Lancaster	78a Penny Street, LA1 1XN	www.eco-action.org/lancaster
4 Oblong	Leeds	47 Westfield Road, LS3 1DG	www.oblongleeds.org.uk
5 Blackcurrent Centre	Northampton	24 St Michael's Avenue, NN1 4JQ	www.ofek.com/bca
6 OARC - Oxford Action Resource Centre	Oxford	Princes St, OX4 1HJ	www.theoarc.org.uk
7 LARC - London Action Resource Centre	London	62 Fieldgate Street, Whitechapel, E1 1ES	www.londonarc.org
8 Synergy Farmers	London	220 Farmers Road, SE5	www.thesynergycentre.com
9 Reading International Solidarity Centre	Reading	35-39 London Street, RG1	admin@risc.org.uk
RADICAL BOOKSHOPS / DISTRIBUTION			
1 Word Power	Edinburgh	43 West Nicolson Street, EH8 9DB	www.word-power.co.uk
2 AK Press	Edinburgh	PO BOX 12766, EH8 9YE	www.akpress.org
3 News From Nowhere	Liverpool	96 Bold Street, L1 4HY	www.newsfromnowhere.org.uk
4 Re-pressed	Leeds	c/o 23-25 Wharf Street, LS2 7EQ	www.re-pressed.org.uk
5 Slendermeans	Sheffield	PO BOX 1251, S11 8XG	www.slendermeans.org.uk
6 Little Thorn Books	Leicester	73 Humberstone Gate, LE1 1WB	
7 Housmans	London	5 Caledonian Road, N1 9DX	www.housmans.com/books/index.htm
8 Freedom	London	84b Whitechapel High Street, E1 7QX	www.freedompress.org.uk
9 Broken Arrow	London	13 Leigham Hall Parade, SW16 1DR	
10 Active	London	BM Active, WC1N 3XX	www.activedistribution.org
11 October Books	Southampton	243 Portswood Road, SO17 2NG	www.octoberbooks.org
12 Natterjack Press	Brighton	mail@natterjackpress.co.uk	www.natterjackpress.co.uk
13 Archway Books	Falmouth	Market Strand, TR 11 3DB	archway@riseup.net
ALTERNATIVE CAFÉS			
1 Forest	Edinburgh	3 Bristo Place, EH1 1EY	www.theforest.org.uk
2 Beanos	Leeds	70 North Street, LS2 7PN	www.beanowholefoods.co.uk
3 The Treehouse	Bradford	2 Ashgrove, Great Horton Road, BD7 1BN,	01274-732354
4 Crocus	Nottingham	Unit 2 Church Square, NG7 1SL	www.thecrocus.co.uk
5 Pogo Café	London	76 Clarence Road, Hackney, E5 8HB	www.pogocafe.co.uk
6 Cafe Kino	Bristol	3 Ninetree Hill, BS1 3SB	www.cafe-kino.com

Figure 13.2   Autonomous spaces in the UK and Ireland

Source: Paul Chatterton and Stuart Hodkinson.

owned. Big civil society convergences outside the summits of global political and business leaders at Cancun, Seattle, Gleneagles or Evian, for example, have seen renewed interest in setting up temporary autonomous villages for discussion, action and putting alternatives into practice. Below, we identify a number of key roles that autonomous spaces can play in building resistance to capitalism.

Directly confronting the logic of capital

The first and most important role of self-managed autonomous spaces is that by reclaiming private property and opening it back up to the public as non-profit, non-commercial zones, they act as a direct ideological and material confrontation to the commodifying logic of capitalism and the process of enclosure. Taking over empty and abandoned buildings like warehouses, factories, garages, schools, shops, clinics, pubs and bars, and turning them into places for politics, meetings and entertainment creates an immediate social and physical barrier to further corporate takeover. Just as the collective organisation of workers into trade unions often forces capital to change course and make concessions on wages and working conditions, the collective reclaiming of public space forces local neoliberal elites to rethink their strategies of gentrifying urban areas by replacing free or cheap places for people to meet and socialise with corporate entertainment, chain stores and luxury flats. For example, V for Housing, V for Victory is a movement in several Spanish cities operating out of autonomous spaces demanding the right to housing for everyone, and under the banner of the 'Popular Assembly for a Dignified Place to Live' in September 2006, 15,000 people took to the streets of Barcelona as a show of strength against speculation and gentrification. Autonomous spaces thus constitute a new claim to how we live – a demand that land and property be used to meet social needs, not to service global, or extra-local, capital.

By challenging the very logic of capital, and the assumed right of the capitalist class to monopolise space, autonomous spaces will inevitably face efforts to repress, shut down and reclaim them. However, when enclosure is directly experienced it can often lead to the radicalisation of whole swathes of people and the creation of confrontational politics to defend what is theirs.

Box 13.1 *ABC No Rio. Social centres East Side style, USA*

Long-term squatting for public spaces is not on the agenda in most of the USA, although there are a few exceptions such as ABC No Rio in New York's Lower East Side. Its origins date to the Real Estate Show, New Year's Day 1980, when 30 artists occupied an abandoned building and mounted an exhibition addressing New York City housing and land use policies. That show was quickly shut down by the police and artwork confiscated. The City was forced into negotiations with the artists and offered them the storefront and basement at 156 Rivington Street. That space became ABC No Rio. It is now collectively run and known as a venue for art activism with gallery space, zine library, darkroom, silkscreening studio and public computer lab. (See www.abcnorio.org)

Creating spaces in which self-organisation, solidarity and mutual aid can flourish

In the individualised, competitive rat race of capitalist society, self-managed autonomous spaces can become environments in which people are genuinely able to relate to and treat each other as equals with solidarity and mutual respect. Core to this is the practice of 'self-management', which rests on a number of key ideas: horizontality (without leaders); informality (no fixed executive roles); open discussion (where everyone can have an equal say); shared labour (no division between thinkers and doers or producers and consumers); and consensus (shared agreement by negotiation). These ideas create a 'DIY politics' where participants create a 'social commons' to rebuild service and welfare provision as the local state retreats. For example, autonomous social centres provide free or by donation meeting spaces, a radical bookshop or library, a cheap cafe, cinema and gig rooms, free shops and internet access. Some will put on computer lessons, benefits advice, language classes, bike workshops and crèches. Several also provide temporary refuge for the homeless, international activists and destitute asylum seekers. But this social commons should not be about recreating traditional public services on the cheap – it is instead about inventing alternative, and parallel, economic models to capitalism based on need, not profit, and respect for the planet.

Putting into practice the horizontal politics of autonomous movements in the everyday, self-managed spaces can contribute to developing a politics based on: the freedom to make our own rules collectively and non-hierarchically without the state telling us what to do; alternative ways of living based on solidarity and mutual aid; and the ability to be who we really are through the rejection of discrimination and domination.

Spaces for uniting social movements, strengthening activism and thinking 'strategically'

As clearly defined places for anti-capitalist organising and non-hierarchical politics, autonomous spaces tend to naturally act as hubs for an array of local campaigns and activists to hold meetings, plan actions, create new networks, publicise their campaigns, produce banners, write pamphlets and raise vital funds to keep going. In fact, many are consciously set up for this purpose as the possibilities for holding free meetings in pubs, community centres and church halls diminish. It is also hoped that such spaces will contribute to strengthening local grassroots movements by bringing people together from different autonomous groups and walks of life in order to create interaction, break down boundaries, create links and community across activist/non-activist divides, and make a locality's autonomous scene into a larger, more coherent and stronger movement. So autonomous spaces can play an important role in reaching out beyond the 'activist ghetto' into the local community and respond to the problems of people in their communities – be it housing, education, property speculators or racism.

Over time, the 'hub' effect often breaks down barriers between existing groups, creates obvious links across what appear to be single issues and introduces new people to radical politics in ways that don't scare them away. Autonomous spaces, and particularly social centres, thus bring the various threads of social struggle together in one place where a process of dialogue, contamination and greater unification can take place. Crucially, these are environments in which people are able to talk through and debate political tactics and campaigning strategies. Autonomous spaces thus create what the Free Association (2005) has called 'safe spaces' for people to retreat to after an intense campaign or large-scale mobilisation; where they can regroup, experiment and enable those going at different speeds and coming from different directions to 'compose together'. If self-managed autonomous spaces can help to link up grassroots struggles, campaigns and activists locally, then they can also connect with each other across cities, towns and communities, and across national boundaries to create a global network of decentralised resistance. Key here is what has been called the 'electronic fabric of struggle' (Cleaver 1998) where the internet can mobilise networks, international solidarity and action.

issues, problems and challenges

We have set out some ideas about how self-managed autonomous spaces can play essential roles in the struggle against global capitalism. As with any political entity, however, these spaces are political experiments that must confront numerous challenges, tensions and even contradictions in their everyday existence. Many simply reflect the reality of practicing radical, self-organised politics simultaneously 'in and against' capitalist society, and the difficulty of putting into practice the values of anti-authority, horizontality and solidarity.

To squat or not?
The biggest and most divisive issue has been the question of 'legality' – to squat or not to squat. Critics suggest that the bureaucracy and paperwork associated with legal autonomous spaces divert a huge investment of activist energy and resources away from grassroots activism and radical social change. Any project or group goes through a process of institutionalisation and professionalisation: activists become

managers of 'social enterprises' or stand in local elections, former squats become licensed and regulated. For many, only through confrontational occupation, not in compliance with the law, will radical politics grow. While squatting is the most risky and short term, it clearly allows people to break free from the constraints and compromises of obeying the rule of private property, avoids paying rent to profiteering landlords and avoids engaging with a system (through licensing laws etc.) they are trying to undermine.

But squatting has its own problems. Some ex-squatters have grown tired, frustrated and burnt out by constantly having to move, finding the very act of squatting itself increasingly difficult in many major cities. Many activists want to put down roots, and make their politics more open, accessible and identifiable. A legal space offers more control. This desire to create more 'stable bases' as a reaction against the ghettoisation of radical/libertarian politics goes hand in hand with the conscious strategic move to create more open and accessible spaces to get people involved in challenging neoliberal policies and strengthen confrontational social movements. Buying or renting private buildings might be far less confrontational than militantly occupying them, but it is often a tactical compromise with the property system towards the same goal of recollectivisation. In any case, all political activity requires some compromises and acceptance from the state, even squatting. Often, squatted and legal places are connected, feeding off each other in a network for autonomous politics.

However, squatting does remain an important act in its own right. In spite of certain dangers and legal implications, squats still need to be undertaken for important political reasons – to directly confront property speculation and evictions, and to quickly raise important issues in the local community or media. Illegality gets things noticed and provokes responses and passions. The Pure Genius land squat along the Thames in central London, for example, raised the issue of how a derelict piece of land was to be turned into a supermarket and car park, while the occupation of the recently evicted Tony's cafe in Hackney, London, was undertaken to highlight rampant corruption of the local council. Squatting is also a vital response to the scandal of empty houses and homelessness in cities across the world. Through  squatting, networks of support and solidarity can often form as people learn and struggle together.

Sustainability

One of the biggest issues in self-managed, not-for-profit projects is that responsibility falls on the shoulders of a small number of people. We all need to take on board the lessons of self-management – that doing it ourselves means doing it together, and that without leaders many more people have to play a role. There's also the issues of how precarious finances often are and how these projects, which don't receive money from the state, often stumble from month to month just being able to pay rent and bills. The reality is that despite their widespread use, often it comes down to a committed handful to actually make that space happen, which can cause problems of burnout, resentment and inefficiency. A particular weakness is the tendency for projects to close during the day as those involved balance their input with other commitments, such as jobs, other forms of activism, family and friends.

Tyranny of structure, tyranny of structurelessness

As living, breathing spaces of relatively unregulated social interactions, autonomous spaces routinely face the challenge of how to put horizontality into practice through structures and participation. Direct democracy often becomes fertile soil for 'the survival of the fittest'. Running such complex entities through mass open meetings can lead to unstructured, draining discussions which are unwelcoming to newcomers. Paradoxically, the desire to avoid specialisation and role hierarchy by making everyone responsible for everything means ultimately no one is accountable for anything, and spaces can rapidly either fall apart as essential tasks remain undone or those with more resources and/or commitment take on more work. What Jo Freeman (1972) coined the 'tyranny of structurelessness', where hierarchies form around those with more experience or knowledge, tempered through the creation of collectives responsible for running certain aspects (bar, cafe, finance, maintenance), but also creating a new layer of bureaucracy (see http://struggle.ws/pdfs/tyranny.pdf). Dealing with incidences of theft, violence or sexual harassment in the space, for example, can also be difficult and time consuming, and requires commitment from the larger group to work together and find solutions. On the flip side it is often the case that groups who have worked together in building a collective to run the space have also examined issues of power and gender relations which are so often sidelined or ignored.

Who, and what, is self-management for?

Many activists involved in self-managed projects wish to 'break out of the activist ghetto' to connect their politics with 'ordinary people' and create welcoming, accessible public spaces attractive to a wide diversity of groups, especially from working-class and ethnic communities. In reality, however, projects are at risk of becoming ghettoised around fairly similar identities (middle class, white, subcultural). At times the aims and ideas of self-management are poorly explained to the public and there is often a breakdown in language. Furthermore, are autonomous spaces really up to the job of creating inspirational and parallel circuits of goods, services and welfare to challenge the profit-driven economy, a centralised state, wage labour, consumer society and environmental devastation? Important problems to face up to include using these projects as springboards into the formal economy, and being recuperated and co-opted into mainstream politics. And what are the implications of the services we provide through self-managed ventures? Are they providers of basic welfare services, filling the gaps the state used to occupy as it retreats, where participants support and offer services to the growing numbers of marginalised people in big cities (the homeless, refugees, migrants)? If this is the case, then such service provision should also inspire us to get involved in managing our own lives and not simply reproduce the dependency culture of welfare services.

future strategies

We have presented a picture of the ongoing story of self-managed spaces and their role in resisting enclosure and dispossession, and reclaiming the commons. Whether they are large social centres, small info shops, ecovillages or protest camps, such spaces continue to mobilise and inspire people to get involved in taking back control of their lives from capitalism. The contribution autonomous spaces make to wider social change, however, is unclear as they remain relatively weak social actors and are largely unconnected from each other and wider civil society. A key priority is building alliances and fighting on more 'bread and butter' issues, such as welfare cuts, job losses, casualisation, housing privatisation, gentrification and hospital closures. It also means developing from being relatively closed 'activist hubs' to open spaces encompassing more traditional social movement actors if broad, popular coalitions against neoliberalism are to emerge. They remain, then, only glimpses of how the future might be, but hold enormous potential to directly confront the logic of capital, create spaces in which human solidarity can flourish, strengthen local activism, pull in greater numbers of people to radical politics and unite grassroots struggles across the world.

Paul Chatterton and Stuart Hodkinson are both involved in the Common Place social centre in Leeds, UK, which since its launch in 2005 has become a vibrant and important self-organised political and cultural hub in the city centre (see www.thecommonplace.org.uk). We are indebted to activists from across the UK social centre movement who told us their histories and perspectives in January 2006.

14 how to set up a self-managed social centre

Matilda Cavallo

Having established in the previous chapter that reclaiming the 'commons' is an urgent and necessary task, how can we put this into practice? How can we, in our daily lives, create space to meet, plan and socialise and reclaim the commons? A quick look around most cities shows us that there is a real shortage of places to live, meet, think, drink or eat, without the government on your back, the corporations branding your experience or your wallet taking a battering. Luxury condos, apartments, air-conditioned offices and chic boutiques are the downtown norm, growing endlessly and attracting those with large incomes whilst at the same time displacing those lower down the social ladder. This chapter shows how we can take back control, if only in a small way, of the places where we live, by focusing on the experiences of setting up independent social centres which bring together a number of activities like meeting spaces, cinemas, cafes or open access computers. These can be applied to self-managed projects of many sizes and types such as info shops, resource centres, community cafes or large occupied communities or land squats. It's not just about what we actually do in these spaces, but also the process of learning and developing our politics we go through to set up and organise these projects.

This chapter provides concrete advice on how to open, organise and maintain these types of self-managed spaces. It combines information for those who aim to squat, rent and buy, which are important political choices. Information in this chapter draws on experiences in the UK and existing guides. While many general ideas remain the same wherever you are, you need to seek out help appropriate to your locality or country, especially in terms of technical and legal matters where differences may have big implications. This is meant to be a guide to help get people started. It is certainly not a blueprint or 'one size fits all' approach – there is a rich tradition of autonomous space projects, all of which provide their own lessons and inspirations.

making a plan

What's the big idea?

The possibilities are endless and there is no one set format, but you need to ask yourself some questions. What do you want to –achieve? What is the actual need? Is there support for the idea and people committed to putting it into practice? Are there groups in the area that would use the place? Have there been squats or other short-term projects, and how did they work out? Will the effort be worth it?

Box 14.1 *To squat or not?*

If it's feasible, don't be put off by squatting to try out your idea. However, be fully aware of the legal consequences of squatting which may involve fines, confrontations with the police, neighbours or other squatters, or criminal records. If you are new to squatting it's definitely a good idea to get involved with people who have experience. The ability to squat takes different forms in different places. In countries like the Netherlands, Belgium and Germany successful negotiations often follow occupations, especially if you can claim residence. In Scotland, Ireland and the USA it is illegal. Squatting in England and Wales is not criminal or illegal, although it is generally seen as such. Thus it is UNLAWFUL, NOT ILLEGAL. For exact laws on squatting and what constitutes 'squatters rights' consult the Squatters Handbook (Advisory Services for Squatters 2005). The main points include:

Squatting: key points

* Find a place that doesn't look too smart – places are often easier to occupy if they are owned by the council, a university or other large public institution. Commercial landlords and property companies are unpredictable
* Get in quietly without doing any damage
* Secure all the entrances and change the lock on the entrance you are using
* Check that the water, gas and electricity are on or can be turned on; sign on for gas and electricity straight away   ▷

> * Make sure that someone is in all the time, especially during the day, at least until the owner or council officials come round
> * If the police, owners or council officials come round don't open the door, but tell them through the letterbox that this is now your home and you are not going to leave until the owners get a possession order to evict you.

What to include?

There are lots of different models you can think about:

★ An info shop which is a focus for a particular campaign or group
★ A radical bookshop
★ A resource centre which offers advice sessions, books and guides, and online resources
★ Independent open-source media centres and hacklabs
★ A cafe or bar which promotes food and drinks politics, such as veganism, fair trading, local sourcing, and organic products
★ Affordable space for gigs for local artists
★ Independent cinema showing documentary films
★ Rural projects, land squats and ecovillages
★ Housing co-operatives
★ Larger centres which bring together many of the above elements.

You also need to ask yourself do you want to buy – which will tie you down for a number of years and start your project off with a gigantic debt, rent – which may tie you down for a few months, or squat – which can be as long as you want, before you get kicked out!

Finding a building

So where do you want to be? In the middle of a local community or housing estate, on a busy high street or in the centre of a city? Each has its pros and cons. You might want to serve a poor neighbourhood or help fight a campaign, or be in a central place with maximising accessibility.

Whether you are looking to squat, buy or rent, the most effective way of finding a building is to walk, bike or drive round every street of the area you are interested in. If you are not sure, note the address and you can do a search for the owner through the Land Registry in the UK (see resources at the end of this chapter).

Box 14.2 Squat spotting

Most areas have large numbers of empty properties. The most important thing is to find a place you are not going to get evicted from quickly. It's difficult to tell how long a squat will last for. It could be a while or you could be evicted in a couple of weeks through the court. It's also important to know if the owner has a record of violent evictions. Opening a new squat is always a bit of a gamble but the more you know the better your situation. Check for signs of someone living or using the building, how easy it is to get in and what the condition of the place is. To find out if someone is going in and out, try sticking a small piece of cellotape across the bottom of the door opening or, better still, a hair held by two small blobs of superglue.

What state is it in? Using our own time and resources we can easily clean up a building. The main thing is that it is structurally OK and that the roof, floors and walls are in good condition.

sorting out the paperwork

1. Renting/buying
If you are going to go down the legal route there will be a lot of paperwork to sort out.

Sorting out the details If you put in an offer to buy a building, you need to get a valuation report and a surveyor's report, building plans, a 'Schedule of Works', and quotes from builders for professional work, and submit these to anyone you are borrowing money from, as well as to Building Control in your council whose approval you'll need. If you are simply renting a space, then you will probably have to do less, but be prepared to get any major building works checked out or done professionally.

If you find a building you want to rent or buy you need to check what type of use it is designated for – and you may need to get a 'change of use consent' from the local government to allow you to do what you want to do. This can take several weeks so don't sign away any money until this is sorted out.

Getting a structure   To run a building legally some kind of registered legal structure will be needed, which means a certain amount of bureaucracy. You will need to appoint a secretary and treasurer, open a bank account, keep good accounts that you have to submit, have regular meetings, and issue members with shares or cards. You can get a bank account aimed specifically at community and not-for-profit groups, from banks such as Unity, Triodos, Ecology or the Co-operative Bank.

Box 14.3 *Types of company structures in the UK*

If you decide to set up a structure for your self-managed space, many people opt for a co-operative company. A co-operative is a business owned and controlled by its members, who also decide collectively on the application of profits. Outsiders cannot have a say in its running and, as a limited company, no one member can be held liable for the co-operative's assets and debts. There are a number of ways you can do this:

* Limited Liability Company registered at Companies House (cheapest and simplest option)
* Co-operative company registered as a Friendly Society
* Community Interest Company (where assets are locked to the community)
* Community Land Trust (for larger land projects).

Making a business plan   A business plan is a plan of who you are, what you want to do and how you will do it. It is what you will be presenting to banks and people you want to borrow money from, so it should be well thought through and fairly respectable! You can get templates easily and they consist of the following:

☆ Basic outline of the proposal
☆ Description of the building and what will happen there
☆ Who will be involved and how it will be organised
☆ A breakdown of all the costs

☆ A cashflow with all projected inflows and outflows

☆ What you want to borrow.

Mortgage and loans If you are buying, look around for ethical lenders and any individuals who may wish to lend you money at low rates over a long period of time. You will usually get a mortgage for about 70 per cent of the cost of the project. As well as a mortgage to pay for the place, remember there will be other costs, such as conveyancing, legal fees, building works, stamp duty and loan repayments until you can generate income.

2. Squatting

The benefit of squatting a building is that you don't have to worry about all the bureaucracy associated with the above. However, even though you don't do so much official paperwork, it's useful to sort out things like collecting and banking money, dealing with solicitors, owners, the media and bailiffs, and getting the utilities turned on.

getting in and setting up

1. Renting/buying

Once you have the building, you've only passed stage one! If you want to open to members or the public, especially if you want to serve alcohol or have entertainment, there are a number of things you need to do.

Getting a premises licence If you want to hold any entertainment that is deemed regulated or sell alcohol legally, then you need to get a premises licence. If you want to operate as a members' club, then you need to get a Club Premises Certificate. It's fairly expensive, there's lots of paperwork, and you will need to meet a list of conditions covering crime and disorder, health and public safety, environmental health, public nuisance and protecting children from harm.

Other things you should consider include: public liability and buildings insurance; business rates (not-for-profit organisations can get discounts); and getting connected to water, electricity, gas and telephone services (where you can use green or ethical suppliers).

2. Squatting

Getting in   The most difficult part of squatting is actually gaining possession, often due to steel doors, window grilles and padlocks. It is illegal to get into a property by breaking in or damaging windows and doors. You could be arrested even if the damage is minimal. You reduce the risk of running into legal problems if you find a property that you can get into easily. Opening a squat by yourself can be risky; it's safer and often more fun to do it with others. Choose a sensible time of the day – most people get a bit jumpy if they hear suspicious noises at night. When you move in, try to make the place look lived in.

> *Box 14.4   Changing the lock*
>
> *The first thing to do after getting in is to change the lock on the front door and secure all the entrances. Until you have control over who comes in and out, you do not have possession and can be evicted straight away if the owner or police come round. It is a good idea for one person to be putting on the lock while others secure windows and other doors, put up curtains, get the kettle on and generally prepare to show that you mean to stay. If there is one, take the old Yale lock off by unscrewing it. Replace the old cylinder with a new one and put the lock back on. Keep the old cylinder in a safe place in case you are accused of theft. You'll need to add a stronger lock, such as a mortice, later on.*

Security   Putting up a legal warning (called a Section 6 Notice, which you can get from the Advisory Service for Squatters or download from their website) in a front window or on the front door may be helpful, as it may deter the police or owner from breaking in. But you must have someone in a place all the time to back it up. A legal warning will not stop you being evicted on its own. Giving a printed copy to the police saves you trying to explain verbally that you can stay if they arrive early on.

Dealing with the police   After you've changed the lock, it is best to start moving your things in as soon as possible. This is the point when the police are most likely to arrive. Don't let them in if you can avoid it. However, the police do have a legal right

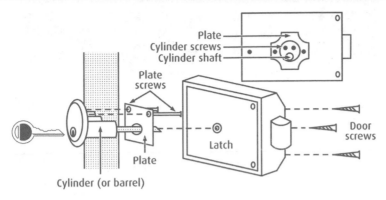

Figure 14.1   Fitting a Yale lock

Source: The Squatters Handbook **(2005).**

to enter a place if they have a warrant. You should tell the police something like: 'We have moved in here because we have nowhere else. We did not break anything when we entered and we have not damaged anything since. It isn't a criminal matter, it's a civil matter between us and the owners, and they must take us to court for a possession order if they want us to leave.'

Getting connected to utilities   Gas and electricity can be deadly. Make sure someone knows what they are doing. If in doubt seek help. If supplies have been disconnected it is probably too much hassle to get the supplies back on. It is an offence to take gas and electricity, and since you can get a supply connected and pay for it, even in a squat, it might be worth it.

organising the space and its activities

The size and dimensions of your space, as well as time and money, shape its possibilities and limitations. Social centres usually become hubs for a huge variety of direct action groups, campaigns and activities including: anarchist, anti-capitalist and working-class politics, peace, anti-consumerist politics, asylum and migration issues, queer politics, co-operative politics, environmental politics, media and creative resistance, international solidarity or anti-privatisation struggles. Whether you are squatting, renting or buying, here are some tips on how spaces can be organised.

Figure 14.2   Fliers from UK social centres

Source: Matilda Cavallo.

Getting people involved

You might not have a shortage of ideas, but it can become a nightmare if it all comes down to a few people working very hard. There needs to be enough people, from the start, interested in taking on different aspects of running the centre, as well as a steady stream of new active members (and there will be a turnover of people) as well as new volunteers. You can't run round forcing people to pitch in, but there are steps you can take to make getting involved as accessible as possible:

★ Have information on how to get involved on membership applications and general leaflets
★ Put up notices around the place
★ Create a rota system that people can sign up to
★ Encourage people to be welcoming to anyone interested in actively helping out
★ Hold 'open days' for those wanting to help run the centre with information on volunteering and talks about the structure of the centre
★ Advertise general meetings well.

How are you going to meet?

Whatever the status of your space, you'll need to set up meetings to make decisions. If you are committed to making decisions without leaders then it's a good idea to set up a way of meeting and making decisions that reflects this (see Chapter 4). Most centres have an open general meeting of all members (weekly, biweekly or monthly) at a set time. Good facilitation is essential to making large meetings run well. Meetings will soon plummet if they are poorly facilitated, dominated by an obvious clique or if people perceive their input isn't valid. Being open and flexible is the key to building vibrant autonomous spaces.

Keeping in touch

Email is an essential organisational tool in today's connected world. But be aware of potential problems and how to resolve them. For example, it's worth remembering that many people still don't have email, so don't use it for making decisions. Make sure you print out any important emails like proposals or agendas and physically post them up on notice boards. Designate someone to make phone calls to contact people who don't have email. Many of us suffer from email overload and if your lists are high traffic you could try different types of lists which people can subscribe to – separating, for example, organising, announcements or political discussions. Emails can also be full of misunderstandings and flaring so make sure you have a clear user guide from the beginning and a good moderator who is prepared to implement it. Rotate this role as it can be a lot of work.

Getting enough resources together

Resources, especially financial ones, are always scarce. Good ways to raise money include membership fees, direct debits, benefit gigs, donations, cafes, bars or bookshops selling solidarity products like T-shirts, fair trade coffee, CDs and DVDs. Some social centres buy buildings big enough to include a housing co-operative, which helps pay towards the running costs of the social centre. But a key resource is enough people willing to support the project in their spare time. Normally, most people will offer their time only for a while leaving a smaller group of volunteers to do the majority of the work. People are not normally paid, which helps towards a feeling of equality.

Getting activities happening

You might have different groups of people already interested in taking on different aspects of the centre. In general meetings you can discuss the potential uses of the space and identify areas that need collectives to figure out what to do. Remember to have clear contact details for all these groups and a way that people can propose new activities and be supported in getting them going. Social centres are about helping people to feel that they are not just consumers. A list of potential collectives includes:

☆ Cafe
☆ Bar
☆ Gigs/events
☆ Bookshop
☆ Garden

☆ Skills sharing
☆ Theatre
☆ Film club/cinema
☆ Bike workshop
☆ Art space

☆ Maintenance/cleaning
☆ Finance
☆ Kids' club
☆ Community outreach
☆ Language classes

Connecting with the outside world

One of your aims may be to connect to groups outside your space. These may include your immediate neighbours (very important if you are squatting), potential volunteers and users of the building, campaign groups, those with certain needs, such as asylum seekers, the homeless, young people, those needing welfare advice, etc. Here are some tips:

☆ Make a clear display area in the space on how to get involved
☆ Use a sandwich board or events board outside the building

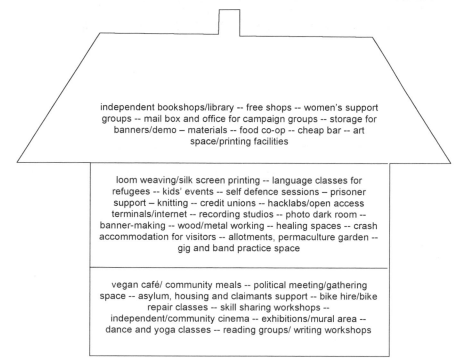

Figure 14.3   Activities in UK social centres

Source: Matilda Cavallo.

★ Make sure working collectives are clearly contactable and have regular, open meetings

★ Set up an email discussion/announcements/organising list but also make sure that those without email are contacted

★ Have a central phone contact or email address that is checked regularly but rotates around different people

★ Have a clear and easy to use website or monthly flier advertising all events happening, how to get involved, how to donate and find the space

★ Work with local groups to target new ones, such as asylum seekers, tenants' groups, those with disabilities

★ Have an attractive leaflet explaining clearly what the space is for and distribute it regularly

★ Make sure information about your space is on relevant contact lists, for example, in the volunteers bureau or on lists of community resources.

Box 14.5   Some key bits of the social centres jigsaw

A members' club
Many social centres set themselves up as members' clubs. This
means access is restricted to members and guests, and you have
to adopt rules covering issues such as joining policy, conduct,
aims, etc. The centre can be open to the public at other times
and the members' restrictions only usually apply when alcohol is
on sale or regulated entertainment is occurring. It's easy to get
over a thousand members just through friends and people visiting.
You can become a members' club by adopting a set of rules at a
general meeting.

Finances
Having a well organised finance collective is a real priority, especially
if there are legal requirements to pay bills on time and produce
accounts.

Cafe
To serve food to the public you will need to be a registered food
business that maintains food hygiene standards. A cafe collective
will need to look into feasible opening times and staffing, a workable
menu, suppliers, organising training in food hygiene, setting prices,
and getting all the necessary equipment together.

Organising events
A building with multiple uses may lead to double bookings and
clashes, and needs some co-ordination of users and events. This is
also something to discuss at a general meeting: what is the building
used for and how is this co-ordinated? A bookings co-ordinating
collective is a good way to do this.

Taking responsibility
Running an open, busy space, especially one that serves alcohol, means
that inevitably there will be conflicts. Make sure that volunteers are
well supported and not left isolated or vulnerable. You can undertake
de-escalation training to help with conflict and set up a mediation
group. It's important that there are designated co-ordinators for
important tasks, such as cashing up and locking up.

leaving and moving on

Self-managed spaces close or move on for a variety of reasons. A campaign may end, group energy may be sapped, finances may go belly up, the landlord may cancel your tenancy. In the case of squats, groups are usually evicted.

If you are serious about fighting an eviction then there are key things you should do:

☆ Get lots of support from everyone you know – the more people the better

☆ Set up a rota so there are people there all the time

☆ Set up a watch to keep an eye out for the police and bailiffs. Remember bailiffs often use physical force, so watch out for this

☆ Agree in advance a timetable, for example, how long are you prepared to stay and resist?

☆ Get people with specific skills – climbers for going on the roof, samba band for making lots of noise, someone to write a press release, people to talk to the local media

☆ Contact the media and get them and the community on your side (stress any aspects favourable to you, for example, is it a David and Goliath battle against developers?). What positive things have you done whilst there which will make people support you?

☆ Video it – it will document any illegal actions by the police and bailiffs and may also be useful evidence later. Make some huge banners to advertise what's happening to the world

☆ If it all goes belly up, make a contingency plan and have somewhere else to go

☆ Be prepared to also build physical defences to resist evictions, such as barricades. Have enough people outside and inside to protect the doors and windows.

This chapter has outlined some of the basic ideas, contacts and skills for setting up autonomous spaces. It's hard work and there is no avoiding the problems – dealing with difficult people, lack of money, free time or resources, and problems with the police and authorities. But it is immensely rewarding to be part of a collective, self-managed project. Go and talk to other people who have done this. Make a plan, think about it, get people excited – but at the end of the day experiment and do it!

Box 14.6 Squat eviction

Unless the owner simply gets their place back because nobody was in, they must apply to the courts for a possession order. Any other method will probably be illegal. Illegal evictions are rare but you should always try to take action against anyone who does this or even tries to. In the UK, the usual claim for a possession order against squatters is a standard one of 'trespass'. But there's the possibility of a nasty optional add-on, an application for an Interim (or temporary) Possession Order (IPO). This was one of the anti-squatting measures in the 1994 Criminal Justice and Public Order Act. The nasty bit is that once an IPO is made against you and served on the squat, normally by letter, it's a criminal offence not to leave within 24 hours, and the police can arrest and charge you. DON'T PANIC ABOUT SUCH A LETTER. It just means the owner has found out you're there. It doesn't necessarily even mean they're going to take you to court soon. That will depend on how urgently they want the place, how efficient they are and luck!

Matilda Cavallo lives between Italy and the UK and has been involved in squatting and setting up social centres for over a decade. This chapter also draws on information taken from The Squatters Handbook *and guides by the Cowley Club social centre in Brighton and the Radical Routes Network in Leeds.*

resources

Books and guides

Advisory Service for Squatters (2005). *The Squatters Handbook*. 12th edition. London: Advisory Service for Squatters.

Cleaver, Harry (1998). 'The Zapatistas and the Electronic Fabric of Struggle'. In *Zapatista! Reinventing Revolution in Mexico*. Eds John Hollway and Eloina Peláez. London: Pluto Press, www.eco.utexas.edu/~hmcleave/zaps.html

Coates, Chris (2000). *Utopia Britannica. British Utopian Experiments 1325–1945*. London: Diggers and Dreamers Publications.

Corr, Anders (1999). *No Trespassing. Squatting, Rent Strikes and Land Struggles Worldwide.* Cambridge, Mass.: South End Press.

Davis, Mike (1990). *City of Quartz.* London: Pimlico.

Davis, Mike (2005). *City of Slums.* London: Verso.

Do or Die (2003). 'Space Invaders. Rants about Radical Space'. *Do or Die* 10: 185–8.

Free Association (2005). *Event Horizon.* Leeds: Free Association, www.nadir.org.uk

Freeman, Jo (1972). *The Tyranny of Structurelessness.* (Pamphlet about informal hierarchies in small groups.) www.struggle.ws/anarchism/pdf/booklets/structurelessness.html.

Hakim, Bey (1991). *The Temporary Autonomous Zone, Ontological Anarchy, Poetic Terrorism.* New York: Autonomedia.

Harvey, D. (2005). *The New Imperialism.* Oxford: Oxford University Press.

Klein, N. (2001). *No Logo.* London: Flamingo.

Midnight Notes Collective (1990). 'The New Enclosures'. *Midnight Notes* 10.

Pilger, John (2002). *The New Rulers of the World.* London: Verso.

Radical Routes (1998). *How to set up a housing co-op.* Leeds: Cornerstone Resource Centre.

Rogue Element (2003). *You Can't Rent Your Way Out of a Social Relationship.* Leeds: Rogue Element.

Shiva, V. (1997). *Biopiracy. The Plunder of Nature and Knowledge.* Totnes, Devon: Green Books.

Stiglitz, J. (2002). *Globalisation and its Discontents.* London: Allen Lane.

The Cowley Club with Radical Routes (2006). *How to Set Up a Social Centre.* Brighton: Radical Routes.

Ward, Colin (2002). *Cotters and Squatters. The Hidden History of Housing.* Nottingham: Five Leaves Publications.

Wates, N. (1980). *Squatting: The Real Story.* London: Bay Leaf Books.

Zukin, Sharon (1992). *The Culture of Cities.* Oxford: Blackwell.

Websites

General

Diggers and Dreamers Guide to Communal Living www.diggersand-dreamers.org.uk

Intentional Communities http://en.wikipedia.org/wiki/Intentional_community

Social Centres http://en.wikipedia.org/wiki/Social_center

The Commoner www.commoner.org.uk/ (A web journal for other values.)

Squatting
Advisory Service for Squatters www.squatter.org.uk
Italian Squats http://tutto.squat.net
No Frills Melbourne Squatters Guide www.geocities.com/squattersguide
Schnews Squatting Guide www.schnews.org.uk/diyguide/squatting.htm
Squat Net http://squat.net
Squatters Handbook Australia www.cat.org.au/housing/book.htm
Wikipedia Definition http://en.wikipedia.org/wiki/Squatting

Networks
Global Infoshops Network www.eco-action.org/infoshops
Italian Social Centres www.ecn.org/presenze
UK Social Centres Network www.socialcentresnetwork.org.uk

Co-operative advice/finance
Catalyst Collective Ltd www.eco-action.org/catalyst
Co-operative and Community Finance www.icof.co.uk
Co-operatives UK www.cooperatives-uk.coop
Industrial Common Ownership Movement www.icof.co.uk/icom
National Community Development Association www.ncdaonline.org
National Confederation of Co-operative Housing www.cch.org.uk
Radical Routes www.radicalroutes.org.uk
Social Centres Network www.socialcentresnetwork.org.uk
Upstart Services Ltd www.upstart.coop
Working Men's Clubs and Institutes Union www.wmciu.org.uk

UK Government sites
Companies House www.companieshouse.gov.uk
Financial Services Authority www.fsa.gov.uk
Land Registry www.landreg.gov.uk
UK Licensing Act 2003 www.opsi.gov.uk/acts/acts2003/20030017.htm

E-lists
Infoshops e-list send an email to lists@tao.ca with the words 'subscribe infoshops'
London Social Centres Network londonscn@yahoo.co.uk
UK Social Centres Network http://lists.riseup.net/www/info/socialcentrenetwork

15 why we need to reclaim the media

Chekov Feeney

On one level the phrase 'the media' simply refers to the various modern technologies for transmitting ideas to large populations, such as newspapers, television, magazines, radio and the new kid on the block, the internet. These are extremely useful tools. They allow people to know what's happening in the world, form opinions and hence share some common (mis)understanding with strangers. This chapter examines the ideas behind what is variously described as alternative, independent or DIY media. This type of media can be differentiated from the mainstream not only through the points of view of those who produce it – although it typically carries a much more radical message – but more importantly by the model in which it operates. It is a model which aims to democratise the process of information production and distribution, a model which aims to allow anybody, regardless of colour, class, gender or how powerful they are, to tell their story and to distribute it to a wide audience. In describing the ideas which underlie this movement, the first question that must be addressed is why do we need such media? What is wrong with the existing mainstream media that moves people to devote time and energy to creating alternatives?

what's wrong with the mainstream model?

The liberal bias model
Sweeping criticisms of the mainstream media are ubiquitous and come from all sides of the political spectrum. Two key, yet differing, criticisms come from the 'right' and from alternative media advocates. Right-wing commentators frequently lambaste the 'liberal bias' of the mainstream media. The 'Media Research Center' is an organisation which describes itself as 'America's Media Watchdog'. It has an annual budget of $6 million and a staff of 60, is mostly funded by corporate donations, and describes its mission as 'documenting, exposing and neutralising liberal media bias'. The criticisms

of such organisations are normally levelled at particular journalists or particular organisations whose reporting does not conform to the right-wing critic's world view. The implicit assumption is that the liberal media is a problem and the biases and prejudices of the individual journalist or her organisation are to blame. These critics tend to express themselves through campaigns to discredit the offending journalists. Thus, for example, Robert Fisk, the British *Independent*'s Middle East correspondent, has been targeted to such an extent by internet critics who find his point of view to be repugnant that the term 'fisking' has entered common internet usage to describe the practice of rebutting an article in minute detail.

These types of criticisms of generalised media bias can be understood as attempts to narrow the range of viewpoints given expression in the mass media. They frequently manifest themselves as campaigns led by vociferous sections of the media, to impose particular points of view and a certain set of assumptions upon opinions expressed in the media.

The propaganda model

By contrast, the critique of the mainstream media that lies behind the alternative media movement rests on an analysis of our modern media at an institutional level. This analysis focuses on the powerful forces which influence the way in which information is conveyed to mass audiences, rather than to individuals. The book and film, *Manufacturing Consent*, by Noam Chomsky (1998) can probably be considered the most thorough exposition of this analysis. It provides a very detailed critique of how news is created and disseminated according to what Chomsky calls the 'propaganda model': a series of information filters which serve to tailor information to the needs of the powerful. Rather than describing these influences as filters, which block out certain information, it would be more accurate to describe them as forces which tend to push the points of view expressed across the entire industry in a particular direction. These forces are not omnipotent by any means, individuals or individual publications can ignore them and react against them, and their effect only becomes fully clear when looking at the media as a whole. They do not operate by issuing edicts, in the manner which the Communist Party used to instruct Pravda, the Soviet Union's state news agency, what to write, but their effect is to skew the overall media output in certain directions. They do not require any conspiracy to maintain them, since the forces are an inherent part of the industry's structure. Nor do they depend upon particular individuals for their operation, since there is no shortage of individuals who will be able to recognise the industry's requirements for a particular slant and fit themselves into the mould.

Chomsky's propaganda model presents, in great detail, a description of the various forces which operate to shape the output of the mainstream media, along with a wealth of empirical evidence to illustrate their effects. The media industry is dominated by enormous corporations and it is largely dependent for its revenue on advertising, much of which comes from other enormous corporations. It is dependent for much of its information on people in important political positions, many of whom have close relationships with these corporations, and on PR agents and lobbyists, who are paid to disseminate the propaganda of these same corporations. The net result is that the media, as a whole, is enormously biased towards presenting a world view that is favourable to these corporations or, more specifically, to those who own the corporations (and their political supporters).

Box 15.1   Concentration of media ownership in the USA

'For better or for worse, our company [The News Corporation Ltd.] is a reflection of my thinking, my character, my values' (Rupert Murdoch, speech to the Asia Society Austral Asia Centre, 8 November 1999).

Six transnational corporations (Disney, Viacom (including CBS), Time Warner, News Corporation, Bertlesman and General Electric) own more than 90 per cent of media holdings in the USA between them, as well as dominating several other markets. In 2005 they had combined revenues of $295 billion and were valued at $550 billion.

There are multiple techniques used to enforce and conceal this bias. Certain sources, drawn from the upper echelons of political and economic life, are automatically considered to be inherently trustworthy, while sources from groups that are in opposition to the powerful are presented as entirely unreliable. Public discussions on important issues are framed as being debates between two poles both of whose positions incorporate the basic bias towards the powerful, excluding the opinions of most of the population from consideration. So, for example, the current debate about Iran has focused on whether 'we' should force them to give up their 'inalienable right' to nuclear power development or whether 'we' should do it by launching a murderous imperial onslaught against them. The various standards and codes of conduct which define the ethics for the industry are themselves dominated by the powerful, and produce a notion of balance and objectivity which merely internalises the bias.

From this point of view the debate between liberal and conservative or left and right forms of mainstream media becomes an illusion. Virtually all mainstream media, liberal and conservative, are controlled by large corporations through advertising and ownership. The media reflects but a tiny slice of opinion and those opinions are limited to points of view which are compatible with the requirements of the powerful. The debates within it merely reflect the tactical differences between different sections of the ruling class and the vast majority of the population is excluded.

For all of these reasons and more, the alternative media movement has attempted to go far beyond the 'the wrong people with the wrong opinions' criticism of the mainstream media. It is an attempt not only to create media which includes a range of voices that are essentially excluded from the mainstream, but also to create alternative structures and even alternative institutions, which do not contain inherent imbalances towards the powerful within them. This movement, if we can call something so diverse a 'movement', is not remotely new.

a brief history of alternative media

Throughout history control over the means of mass communication has been a crucial means of exercising control over society. Technological advances have, from time to time, radically democratised access to mass communication, a democratisation whose echoes have been felt throughout society. In the Middle Ages in Europe the church was able to maintain a firm grip on the social order, largely through its monopolisation of written media. No other institution could compete with the church's network of monasteries and scribes. Only those works which met the church's approval were copied and distributed widely. The invention of the printing press finally broke this monopoly and, with time, this technology became available to wider strata in society. The revolutions that shook the world in the late eighteenth century were heavily indebted to the relatively widespread availability of printed tracts, such as Thomas Paine's *The Rights of Man* (1791), which were crucial in popularising republican ideas. Throughout the nineteenth and early twentieth centuries, the workers' press flourished all around the world – a printing press was until fairly recently the first target acquisition of radical political groups. In many countries, the organised workers' movement could compete on an almost equal footing with the capitalist controlled media. For example, in 1930s Spain the anarcho-syndicalist CNT produced over 30 daily newspapers, including the national best-seller.

However, the second half of the twentieth century saw the precipitous decline of the once flourishing workers' press. On the one hand, the catastrophic pyrrhic victory of authoritarian socialism in Russia gave the world a perfect proof that just because the media might be anti-capitalist doesn't mean that it can't be worse. Pravda became a watchword for the political ruling class exercising total control over mass communication. The fortunes of the workers' press in the 'free world' were similarly poor. Although this reflected, to some extent, the declining fortunes of the workers' movement, that decline was also a consequence of changes in the economics and technology of how media was produced.

The victory of the corporations

New media technologies, such as television and radio, that were introduced in the twentieth century tended to be even more tightly controlled by governments and large corporations as they required greater capital investment. While radical organisations could make up for lack of resources for publishing newspapers through voluntary labour and distribution networks, the capital investment required to construct television networks was largely beyond them. The increasing size of media corporations and their increasing dependence on advertising revenue for income slanted the field further still in favour of the corporations. Not only did publications which carried view points that were opposed to the views of the powerful attract less advertising, but advertisers organised boycotts against such publications. Writing recently on the fiftieth anniversary of the Suez crisis, *Guardian* editor Alan Rusbridger described the 'long-lasting and debilitating' effects of an advertising boycott against the *Observer* newspaper for having reported truthfully on how the crisis had been manufactured by the British and French governments as a pretext to invade Egypt. Thus by the 1980s, after several decades in which these trends had strengthened, most of the formerly flourishing alternative media had been worn down by commercial pressures, sold out or given up. It seemed as if the future of media would be one of ever larger conglomerates with ever stronger commercial and political imperatives driving their content. On the fringes, with miniscule resources, tiny circulations and no pretensions to challenge the mainstream, groups with alternative points of view would put out newspapers, fliers, pamphlets and DIY zines, and would limit their ambitions for mass communication to a small and dedicated following, and the occasional sympathetic voice in the mainstream.

The internet and mass communication

However, as technological advances in the early twentieth century were responsible to a certain extent for creating the barrier to entry which gave such an advantage to the corporations, the advances since the 1980s have had the opposite effect. Cheap personal computers, media production software, digital photography and recording equipment have between them massively cut the amount of capital investment required to produce media that can aspire to compete with the corporations. Most importantly of all, widespread internet access has hugely reduced the cost of distributing media. In 1980 somebody who wanted to produce and edit a short film and distribute it to an audience around the world would have needed access to millions of dollars worth of equipment. Today the same task can be achieved with a cheap digital video camera and a standard internet-connected PC running inexpensive or free software.

In addition to its low financial barrier to entry and its transnational, geographical distance-collapsing nature, perhaps the most important development of the internet is a consequence of its fundamental communication paradigm. Traditional media facilitate 'few-to-many communication'. This means that a relatively small number of people produce the information, while a large number of people consume it and there is a clear division between the two. This model is favoured when there is a relatively high cost involved in producing and distributing the information. In the early years of the internet, this was the predominant model for websites, with sites being managed by individuals and small groups and passively consumed by viewers.

However, unlike a newspaper or a TV broadcast, there is virtually no cost involved in adding and distributing new information on the internet. There are few of the same constraints on the size and volume of the information distributed. This feature has facilitated the development of 'many-to-many communication' models, sources of information created by participatory, voluntary communities where the lines between consumer and producer are blurred. This type of community stretches back to the birth of the internet and has migrated through the various internet communication tools from Usenet newsgroups to email lists, bulletin boards, forums, community driven news sites and blogs on the World Wide Web.

Probably the most impressive child of the internet is the free software movement, a vast and nebulous community of computer programmers, spread all over the globe, who use a production model that is much closer to pure communism than to capitalism – the vast majority of work is voluntary and the products are given away for free. This community is responsible for much of the software that runs the internet itself and its creations have been crucial in the development of internet communities where information rather than software is the product. With the development of software tools to facilitate the creation and distribution of information by large groups of co-operating people, enormous repositories of information have been developed by ever growing communities. The increasing sophistication and ease of use of the tools has been closely followed by larger, more diverse and more sophisticated examples of community organisation.

the indymedia model

Although there are many interesting examples of alternative models for producing and distributing information on the internet, the remainder of this chapter will focus on the Indymedia network for a number of reasons. Firstly, the author has been involved in the network for the last five years as an editor of the Irish Indymedia website and as one of the developers of the oscailt content management system. Secondly, Indymedia is a project which was based upon a radical analysis of the failings of mainstream media and one that consciously attempted to come up with an alternative organisational model in order to avoid being influenced by the various forces which skew the output of mainstream media. Finally, it has always had ambitions to challenge the mainstream media's dominance and, unlike many internet-based projects, it has not generally contented itself with speaking to a niche audience.

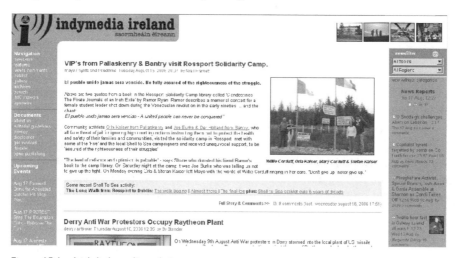

Figure 15.1   Irish Indymedia website

Source: Indymedia Ireland.

Indymedia was born in Seattle in November 1999, during the now infamous protests against the World Trade Organisation and has remained heavily influenced by the radical libertarian ideas current in the global justice movement. It was initially composed of two basic elements: a physical media centre, where social justice activists who were protesting against the WTO could come together and share information, and a website, which anybody could publish stories on, and upload video and audio segments to, as well as add comments to the stories and videos. It proved an instant success. Within a few days it had attracted over a million hits (which was a lot back in 1999) and the idea spread like wildfire. As the 'anti-globalisation' protest movement spread around the world, Indymedia sites followed in the footsteps of the protests. Groups all over the world came together to set up their own local version of the Indymedia site, based upon one of the freely distributed open source content management systems written by Indymedia activist programmers. Indymedia collectives branched out to establish radio stations, video production groups, newsletters and a wide variety of alternative media offerings.

Figure 15.2   Reclaim the media logo

Source: Indymedia.

Today, Indymedia has expanded to become a global network of open publishing news sites, with over 150 collectives of varying size in over 70 countries. 'Open publishing' means that all of the users of the site produce the news collectively, rather than it being a job of a small group. The members of each collective are responsible for enforcing basic editorial guidelines and choosing which articles to highlight as 'features'. The network of collectives agrees to a basic set of goals and principles as part of the process of joining. These network wide agreements amount to a statement of basic anarchist organisational principles – emphasising democracy, equality, accountability, openness and non-hierarchical structures. They also emphasise, in contrast to the mainstream media, the fact that they do not intend to present news in an objective or balanced manner. The network's basic 'about us' page declares that Indymedia should be about 'radical, passionate tellings of truth'. The idea is to promote accuracy rather than objectivity and to allow all sides to tell their own version of events, so that a richer and more nuanced picture can emerge from the whole. However, beyond the basic agreement of principles, the collectives are autonomous and have great lassitude to interpret the guidelines in different ways.

The Indymedia collective is held together by a collectively managed technical infrastructure which comprises hundreds of mailing lists, internet chat channels, occasional real world conferences and a variety of 'syndication' sites, which pool together news on an issue-by-issue or regional basis.

Although many people thought that the Indymedia principles of non-hierarchy, open publishing and consensus decision making would, by themselves, solve many of

Box 15.2 Different approaches to Indymedia

Different Indymedia sites in different countries have taken markedly different paths over the years. For example, although they are close together Indymedia UK and Indymedia Ireland have some important differences:

* Organisation: Indymedia UK is organised as a network of regional collectives spread around the UK. Each regional collective meets and makes decisions locally which affect their regional pages and local collectives. Network wide issues are decided on mailing lists and at occasional regional meetings. Indymedia Ireland is a single collective which makes almost all of its decisions on mailing lists.

* Editorial: Indymedia UK generally allows material to be cross-posted to their site as well as to other sites, while Indymedia Ireland only allows original content. Indymedia UK does not allow postings from explicitly hierarchical groups, while Indymedia Ireland allows postings from all political currents, including right-wing parties. The only limit on political content is a ban on discriminatory or hateful postings.

* Comment editing: Comments and articles that are against the guidelines on Indymedia UK are moved to a hidden page, while on the Irish site they are removed altogether and can only be viewed by subscribing to a special mailing list which records all editorial actions.

the problems that beset the mainstream media, this turned out to be naïvely optimistic. The first and most immediate problem that most Indymedia collectives faced was to do with the thorny question of 'censorship'. Many inexperienced volunteers had imagined that if a news site allowed people to publish whatever they wanted without any censorship at all, this would eventually lead to the more coherent and better reasoned points of view eventually winning out. The early years of the Indymedia network saw fierce debates between the advocates of absolute free speech and those who advocated some form of content selection and removal. In the end the argument was won not by arguments or reasoning but by reality. The cost of reproducing information on the internet is virtually zero and it is trivial for somebody with destructive intent to swamp an open communication channel with disruptive content. Sites which adopted a free speech absolutist position quickly found themselves engulfed with right-wing trolls, neo-Nazis, spam and anti-social lunatics. Collectives that tried to argue with these abusers rather than ban them eventually ran out of energy – there's only so many times that somebody will bother to refute the same stereotyped propaganda before they give up.

Thus, those who advocated some form of content filtering eventually won by default as genuine users stayed away from the rubbish filled newswires of the free speech sites and collective members eventually burnt out. Still, although there was a general acceptance about the need for some filtering of content on the newswires, there was no widespread agreement as to how such filtering should be carried out. As the project is fundamentally an attempt to organise media production without any inherent biases and without a hierarchy of individuals imposing their agendas on the public, the question of filtering information remains a contentious issue in Indymedia. Some sites adopted systems where users needed to register before they could post to the site. Others essentially decided to ban content with a right-wing angle, reasoning that such content had more than enough channels for distribution already. Others moved towards 'professionalisation' whereby skilled editorial volunteers or paid staff would verify the facts of submitted stories before publishing them to the prominent newswire. Others adopted increasingly strict guidelines defining the requirements for newsworthiness to enable material to be published on the site. Most sit somewhere in between, removing disruptive content and personalised abuse, but allowing input from all political points of view as long as they do not contain hate-speech, such as blatant racism, sexism or homophobia.

The other major problem that the Indymedia network has grappled with is the issue of collective decision making. The momentous early growth of the network and its firm commitment towards consensus decision making quickly created a situation where collective decisions became impossible to reach. Trying to get several hundred collectives, with dozens of different native languages, to unanimously agree on any particular decision is basically impossible. This has led to a situation where network-wide decision making has become impractical and many collectives are, in practice, entirely autonomous with some only bearing a shallow resemblance to the stated aims and principles. For example, despite the fact that a large number of Indymedia volunteers have declared that the Belgian collective is controlled by the Stalinist-Maoist Belgian Workers, meaning it therefore does not conform to Indymedia principles, as well as the lack of convincing refutations, moves to remove the Belgian Indymedia site from the network were repeatedly blocked in 2005.

The inherent looseness of the collective structures and the overall network has also caused several sites to fail to deal with the relative downturn in activity among the global anti-capitalist protest movement. Without permanent structures to support them, many sites have melted away as the energy of the small number of volunteers who sustained them has waned. However, the picture is not all bleak. A considerable number of Indymedia sites, particularly those from southern Europe and South America, have managed to broaden their audience and have continued to grow despite the relative decline of the summit protests which were once their core subject.

Going forward, there are a number of lessons that can be drawn from the evolution of Indymedia. We should not ignore its enormous success – starting from nothing with no resources and entirely dependent on volunteer labour, it managed to spread around the world and distribute alternative points of view on a mass scale. However, we should also not ignore the problems. The relative looseness of the organisation and the naïve belief that an open, unfiltered news service would be of much value caused an awful lot of time, energy and enthusiasm to be wasted. More structured co-ordination between collectives, with democratic, accountable and objectively applied editorial criteria on a variety of levels, could see the Indymedia network become greater than the sum of its parts and if it succeeded in that, the corporate media would have real problems.

Jumping hurdles

There are also a number of hurdles that the Indymedia network will meet in the coming years. For a start, there is always a danger that any movement can stagnate and settle into a comfortable niche. Alternative media projects, which rely upon volunteer labour, always have to deal with the fact that people run out of time and energy, and drift away. It is vital that Indymedia collectives continue to attract new members and seek out new audiences. The other side of the coin is that there is always a risk that a collective can lose its radical ethos and become incorporated back into the mainstream system. For example, many formerly radical community radio stations have, over time, become dependent on government, commercial or NGO funding which has inevitably eventually extinguished the radical ethos which marked them out. Turning away from the aspiration for open access and towards professionalisation also carries attendant risks – the project becomes the vehicle for the points of view of the members of the group and although these points of view may be radical, the radically different model of information production is diluted.

Another potential hurdle is the issue of repression. For many activists and people involved in social movements across the world, their first port of call for reporting an action or event is Indymedia. The fact that you can self-publish, access it 24 hours a day, find out news that you won't read in the papers, and comment on articles and events means that Indymedia has become a crucial and vital tool. Success breeds contempt and there are many instances of Indymedia sites, centres and journalists being deliberately targeted and attacked by the authorities. On 7 October 2004, the FBI seized some of Indymedia's servers, hosted by a US-based company. The servers in question were located in the UK and managed by the British arm of the company, but some 20, mainly European, Indymedia websites were affected, and several unrelated ones (including the website of a Linux distribution). No reasons were given at first by the FBI for the seizure. In June 2005 a member of Bristol Indymedia was arrested by police in the run-up to the G8 summit in Scotland and charged with incitement to criminal damage after an anonymous person published news of an action involving property damage. The page had been promptly hidden by editors of Bristol Indymedia. Despite being advised by lawyers of the illegality of such action, the police still seized the server seeking access logs from the server operators. The incident ended several months later with no charges being brought by the police and the equipment was returned. Attacks on Indymedia journalists and Indymedia's centres have also occurred in the past. One of the most brutal was in 2001 during protests against the G8 in Genoa when police stormed the DIAZ school that was serving as an Indymedia centre at night. Video evidence by people that were able to hide and film the attack showed police brutally beating sleeping people. Twenty-nine Italian police officers have since been indicted for grievous bodily harm, planting evidence and wrongful arrest, and a further 48 officials have been charged with torturing activists and journalists that were arrested in the raid. The court case is still ongoing. In 2005, a US arms manufacturer, EDO-MBM, threatened Indymedia with libel action after a series of articles called the EDO-MBM (UK) company 'warmongers' for selling arms to Israel and the US forces. The company never saw the action through. The strong network of Indymedia and support it receives for its work means that for the most part incidents of repression against Indymedia are well documented, reported and challenged. Yet as Indymedia grows in success and becomes a well recognised and credible source of information, the possibility of facing law suits and further server seizures also increases.

towards a new media

Although this chapter has focused primarily on Indymedia, there are many other projects in other areas which share a lot of similarities with its goals and networks. For example, there are community radio stations, radical video production groups, alternative news print publications and even public access television stations scattered around the world which all aim to provide democratic access to the media. To provide a few very brief examples: 3CR is an entirely listener-supported community radio station in Melbourne, Australia, which provides space to a wide range of alternative and radical voices; Undercurrents is a UK-based radical film production and distribution group which focuses on producing documentaries highlighting social justice issues and radical protest movements; and Schnews is a free weekly direct action news sheet from Brighton, UK, which has produced many highly acclaimed books, pamphlets and films in addition to their weekly news sheet. These are but a tiny sampling of the alternative media projects out there, but they are distinguished by the fact that they have all survived and thrived for over a decade, reached a wide audience and managed to retain their radical politics.

These projects show what is possible when people get together and put their ideas into practice. In the age of accessibility, the potential of creating our own media, shouting with our own voices, telling our own stories can be realised. We don't have to rely on the moguls to tell us about the world. We can get our stories from our neighbours and from other people in struggles all over the world. We have the power to describe the world as it is and we can put aside the ideological blinkers which power would put over our eyes. We don't have to hate the media, we can be the media.

Chekov Feeney is a political activist and journalist living in Dublin. He is part of the Irish Indymedia collective (www.indymedia.ie) and writes a regular column for the weekly Village Magazine *(www.villagemagazine.ie).*

16 how to communicate beyond tv

Mick Fuzz

media activism as a tool for building communities

We live in a highly mediated age. We've been brought up on well edited and engaging text and images. Be it print, radio or video – media is a craft tailored to be convincing. It makes it easy to mutate reality. If we don't tell our own stories then we risk leaving others to create conflicting versions which will be the only ones that resonate. Building our own media has become increasingly easy with recent technologies. Free software such as Linux and Mozilla can be seen to embody mutual aid, WIKI and open posting websites break down hierarchies of communication, Indymedia allows everyone to make their own news, while blogs (web logs) encourage reflection of our place in the world and free expression. Collaborative media projects can also be a good way of building trust and working together on a unified aim and are an action in themselves. This chapter looks at a few different tools (community newsletters, spoof newspapers, participatory video, community film screenings) that are used regularly by people to create their own media.

how to set up a community newsletter

Starting up a local newsletter is something that can get lots of people together working collectively and learning about local issues. The guide below is adapted from the *PorkBolter*, a successful radical news sheet based in Worthing in the UK.

Stage one: Organise a meeting
You've talked about it down the pub with a few mates. You all think it's a great idea. So just get on with it. Fix a date, time and venue. Leave other possibilities wide open. It's important for everyone to have had a say in the shaping of the project from the start.

Stage two: Get it all sorted

Make sure you agree some things – a name and address, which will in turn enable you to set up a bank account in your newsletter's name. For the name, the main requirements are that it should have a local reference and that it shouldn't be too overtly political – you are addressing ordinary people.

Stage three: The details

You've also got to start thinking about boring details, like size, frequency, number and so on. You'd be amazed at how much you can fit on a double-sided piece of paper. As far as frequency is concerned, once a month seems about right for many. Quantity is obviously limited by funds. Try getting 500 done to start with, then 1000 if your distribution is working.

Stage four: Printing

Cheap photocopying or printing is hard to come by, but very useful. Don't just rush out to the nearest High Street print shop. Try your local student union or college print department or local resource centre. If all else fails, appeal to readers for information.

Stage five: Paying for it

You'll probably find yourselves fulfilling this role. But spread between the group members it doesn't come to much. If you meet at someone's home instead of in the pub, you'll probably have saved enough for the next issue. Other costs may well be covered by donations, fundraisers or subscriptions.

Stage six: Getting it out

Distribution is easy when it's free. It's just a question of getting the newsletters all out locally. You can do that by standing in the town centre and thrusting them into people's hands or leaving them in the library and town hall, and in shops and pubs. People should also be able to subscribe for a small charge to cover postage.

Stage seven: Contents

What do you put in the bloody thing? First of all you read all the mainstream local papers. And then you get very angry with all the stuff the council's up to. Cut out the relevant bits and bring them along to the next newsletter meeting. Someone writes down the best bits and the contents start to emerge. Add in your own little campaigns (anti-GM, anti-CCTV, anti-negative attitudes, etc.) plus titbits about worthy local groups and you've got a newsletter.

Stage eight: Campaigns and keeping it local
Trying to persuade people that global capitalism is a bad thing because it is destroying the Amazon rainforests is a waste of time. But talk to people about the way that money-grabbing property developers are allowed to build all over green spaces on the edge of your town and your readers will understand why you call for an end to the rule of greed and money over people. In your newsletter your views can clearly be seen as common sense. You are normal and the council or property developers or government are the outsiders – reversing the way radical views are conventionally presented.

Stage nine: Have a laugh
A jokey approach makes people read your newsletter and explodes certain ill-founded stereotypes about radical political initiatives. Could be a problem, though, if your group does in fact happen to be entirely composed of humourless left-wing gits.

Stage ten: Law abiding
Remember that you can get done for libel if you make certain claims about individuals. Get round this with humourous digs and heavy use of satire and sarcasm (think *Private Eye*, Michael Moore). It is worth knowing that you cannot libel a council – so go for it!

Stage eleven: Carry on publishing
There will be ups and downs. New people will join your circle. Others will drift away. It might seem like nobody's taking any notice of you at all. But in fact your message will be permeating the very fabric of your community. It's got to be worth it.

spoof newspapers

Since April 1997 when a 20,000 print run of *Evading Standards*, a skit on London's *Evening Standard* was published for a march in support of the Liverpool dockers, the mockery idea has spawned its own imitators worldwide. That original issue was seized in bulk by police and three people were charged with incitement to affray. But the spoofers sued the Metropolitan Police for wrongful arrest and were awarded five-figure costs. The money funded the next edition, and *Evading Standards* made a return for a demonstration in the City of London. Publishing alternative newspapers is an

Figure 16.1   *Bristle*: Bristol's local monthly magazine

Box 16.1   *The Hate Mail newspaper*

The Hate Mail was created by the Manchester No Borders Group. Our aim with this project was to create a publication that contained our views on immigration law in an accessible format. We wanted something that we could give out to the general public about the violence against and inhumane treatment of people in the asylum system. We wanted to help widen the spectrum of acceptable debate on immigration issues. After all, campaigning to achieve leave to stay on compassionate grounds is not a radical stance.

Together we worked out the ideas we wanted to communicate and divided them up. When the group was happy with the ideas we set a word limit for each article and we went away to complete it. Most main articles were written with some kind of image in mind. The artwork and layout and the final editing were specialised roles taken on by one or two people. A tabloid publication is relatively cheap per issue to print when you get above a certain number. It is suitable when you have a message that you want to get out to tens of thousands of people. You need a network of volunteers that are prepared to go and give it out. In this case we contacted members of autonomous social centres, resource centres and housing co-ops and found out how many hundred they would commit to giving out before printing.

Tabloid coverage of some issues relating to asylum and refugees is so extreme that to create a spoof of it became quite difficult. So we made the decision to include dark and twisted humour designed to shock. The front cover headline read: 'Asylum Seekers Ate My Hamster'. We decided to make the first four pages funny before hitting home with our real message, an editorial, a 'how you can help' resource page and a myth-busting section called 'You are being lied to about asylum seekers'. The editorial of the Hate Mail was as extremist and simplistic as tabloid editorials. I.e.:

WE KNOW that this island is blighted by irresponsible media reporting that has a TERRIBLE and lasting impact on community relations... WE BELIEVE in freedom of movement for all people on this planet. We believe that all humans have the right to decide where they want to live and who to fall in love with.

attempt by campaigners to bypass the mainstream media and deliver their message directly by imitating the style and layout of well known and recognisable newpapers. Other recent spoofs have included *Shitty Life*, the *Newcastle Evening Chronic* and the *Financial Crimes*.

When the concept crossed the Atlantic, the game moved on. Protesters against the World Trade Organisation produced a quality copy of the awkwardly named *Seattle Post-Intelligencer*. Squads of early morning paperboys and girls simply opened up the street boxes with a quarter, replaced the outer four pages and put them back. They were attacked in an angry editorial the next day. The *San Francisco Chomical* used the same technique to highlight the incarceration on death row of black activist and writer Mumia Abu-Jamal.

One such spoof newspaper that was circulated in the UK in 2005 was called the *Hate Mail*. This paper spoofed the often misleading coverage of refugees and asylum seekers by the right-wing press in the UK.

Creating a spoof newspaper – checklist:

✓ Decide on the communication objective: What is the theme and message?

✓ Make sure it is clear who the target audience is: If you're speaking to kids, then your language and arguments will have to be understandable to kids.

✓ Decide on format: Is it going to be a magazine, newspaper or leaflet? Make this decision based on the target audience you're trying to reach and the amount of money available.

✓ Decide on concept: The concept is the underlying creative idea that drives your message.

✓ The visuals: Though you don't absolutely require a visual, it will help draw attention to your ad. Research indicates that 70 per cent of people will only look at the visual in an ad, whereas only 30 per cent will read the headline.

✓ The headline: Your headline must be short, snappy and must touch the people that read it. Your headline must affect the reader emotionally, either by making them laugh, angry, curious or think. If you can't think of a headline that does one of these four things, then keep thinking.

✓ Distribution: Decide on distribution channels and make sure they are ready to go. Be aware that the police have in the past seized spoof papers. It may be worth dividing the papers up around different locations.

Hate Mail

IMMIGRATION PRE-ELECTION SPECIAL   www.makebordershistory.org   National Edition

Dirty Tricks, Propaganda and Outright Lies
See page 5 for details

THE NASTIER THE BETTER!
Bidding war on asylum escalates

In the past we've brought you many unbelievable stories about foreigners assaulting the British way of life - today is no exception.

ASYLUM SEEKER ATE MY HAMSTER

By Felix Morass

Reports have come in that send the level of asylum lunacy to new heights of ear-bursting irritation. Our sources have revealed that ever-hungry border-hoppers have been eating hamsters. Animal lovers and medical experts are equally outraged. The British hamster is rightly renowned as one of the finest examples of pedigree rodents world-

Have generations of endearingly eccentric English pet lovers created a truly pedigree species only for it to be scoffed by benefit cheats?

These gorgeous mini thoroughbreds, sum up what it is to be a proud member of the Royal Realm. But now, the loveable creatures are at risk from the human vermin who, day-by-day, gnaw away at the skirting boards of decency.

Full story and editorial - page 5...

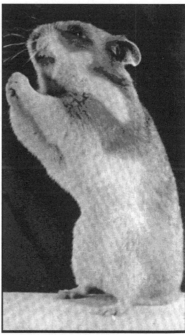

A terrified 'at risk' rodent pleads for its life, yesterday.

INSIDE: You Are Being Lied To 6, Hate Apathy 4, Hate Opinion 5, Hate Life 9, Hate TV 10, Hate Yourself

Figure 16.2   The *Hate Mail* spoof newspaper

participatory video

You come with preconceived shots and ask me 'what are you doing today? I want to film you'. I get worried, not knowing what happens to what I say. I speak in my language; you write in your language and make your film in your language. If I see your film, I feel, I never meant this. Only you educated people have the chance to own and control media. You advertise sprayers, pesticide, toothbrushes and fertilizers. Whatever you produce, you have the right to show as long as you desire. When we, the hardworking people, want a media of our own, you tell us that we cannot have one. We also want to show our issues and problems. Some of you must start thinking about us and give us the right to own our own media. (Chinna Narsamma, a *dalit* woman in the village of Pastapur, Andra Pradesh, India, explaining why she and her group now make their own videos)

Participatory video (PV) is a term that was coined in the 1960s alongside Paulo Freire's popular education. The point is to give a voice to people whose voices would not always be heard. PV could be defined as a scriptless video production process, directed by a group from the grassroots. It aims at creating video narratives that communicate what participants really want to communicate. Rather than making a documentary about an issue, the people directly involved in the situation make the video about themselves – from beginning to end. Participatory video has been used to great effect to give voice to the marginalised and unheard.

Participatory video in a nutshell

* ★ Participants rapidly learn how to use video equipment through games and exercises
* ★ Facilitators help groups identify and analyse important issues in their community
* ★ Short videos and messages are directed and filmed by participants
* ★ Footage is shared with the wider community at screenings.
* ★ It's an excellent tool for public consultation, advocacy, policy dialogue, for mediation in conflicts, development projects and programmes
* ★ It can be used to communicate the outcome of all kinds of participatory assessments and processes, in all stages of the project cycle – planning, monitoring and evaluation

☆ It can be applied horizontally, for communicating between local communities, and vertically, for grassroots people communicating with policy makers, donors and government authorities

☆ With the help of voice-over translation it can be used to communicate across languages

☆ Where there is local television, PV can extend a local process of planning or consultation to wider populations.

Participatory interaction has several key objectives in building a video: to show people how to use a video camera; to learn video techniques; to build confidence both behind the scenes and in front of the camera; to build group dynamics; to explore different issues; and to learn to tell stories. The game below from Insight Video can be used as part of a participatory video training programme.

Box 16.2   The disappearing game

Objectives: Have fun, group building, learn how to record and pause.
Numbers: 3+.
Duration: 10–20 minutes.
Materials: Video camera, TV monitor, tripod, AV (audio-visual) lead.
Stages

(a)   The whole group of participants stands in a group as if posing for a photograph.

(b)   Person A is filming and should ask the others to stand like statues and be silent. Try to make it funny; for example, standing in silly poses.

(c)   Person A pushes the button and counts to three (records for 3 seconds). If the camera or tripod is moved, even slightly, the trick will be spoiled.

(d)   Person A asks someone to leave the group – remember the others must not move.

(e)   That new person then pushes record.

(f)   When the last person is removed, film the empty space for 5 seconds.

(g)   Now watch it immediately. It will look as if people appear and disappear as if by magic.

Box 16.3 *How to storyboard*

Storyboarding is a way of setting out what is going to happen in your video. It's useful to agree on the main theme or story before starting the storyboard.

Objectives: Develop participants' confidence and control over the process, build group working skills, share roles, learn to tell a story with images.
Numbers: 3+.
Duration: 1–3 hours.
Materials: Something to draw on, video camera, tripod, microphone, TV, AV lead.
Stages

(a) Talk to participants – find out what story they would like to tell. You can use creative activities to stimulate ideas if necessary. Ask them: 'What would you like to make a short film about?' Build their confidence, encourage and praise their ideas.
(b) Draw 4–6 boxes.
(c) Ask: 'How would you introduce your story?' Draw a sketch in the first box. Draw a simple image (stick figures, for example).
(d) Continue quite rapidly with the outline story – try to get participants to draw in the boxes themselves. Make sure everyone is involved.
(e) At the end go back and get details for every box: 'Who is talking here?', 'Who is filming this shot?', 'Where will you film it?'.
(f) Congratulate them.
(g) The group now goes to film the shots in the order laid out in the storyboard.

community film screenings

Films screenings are a great way of getting together in the same room with a common purpose. There are plenty of reasons for creating a forum of people:

✮ To outreach for existing campaigns and projects and get more people involved

✮ Air concerns and bring out debates on issues that affect a community

✮ Informal mediation and conflict prevention

✮ Create an atmosphere of inspiration and positive creativity.

Going beyond TV

Ideally video should be just one medium with which you engage with the audience. Live interviews, popular education, spoken word, audience comments, music, requests, heckling and corrections, announcements and earnest calls to action are the real flesh of a film screening night, and the fact that you are showing films just helps facilitate that. You can treat a screening as a show no matter how small it is. If you have put out a film screening flier and told people that it is about social change, you can see it as a duty to make it upbeat, intelligent and likely to make people active. You may be able to do that in the way films are introduced and the way people are asked to make announcements. People notice the spirit in which it is done and if you really care. Often that can be more influential than the content of the films themselves.

At screenings you can show and say things that just don't get on mainstream TV. Here are some things that TV can't do that public screenings can:

✮ Use video flexibly as a resource to stop the action to allow people to share emotions, and reactions, and form a response

✮ Allow its viewers to make announcements of upcoming events as the screening progresses

✮ Inspire people to become active in campaigns and action

✮ Bring out debate in situations where it wouldn't normally happen.

How to put on a successful screening

Choose a venue with an audience in mind. These can include:

✮ Arts venues: They often have good technical support and the sound quality is good.

✮ Club/dance parties: Sometimes doing social justice films in these types of places can get lost in the chaos. If you keep your message clear, perform it well, and have really upbeat comedy and music based films it can work really well.

✮ Pub/bar/social club: There are certain videos that are so strong that they can compete with background noise, and work in pubs and bars. It's good to have a microphone and a pretty good PA, as people chatting in pubs make a surprisingly loud noise.

★ Community centres/churches: These venues can be a bit cold. Try and create a warm vibe somehow.

★ Pirate TV: Pirate TV stations in Italy download programmes from the internet and retransmit them via the airwaves. One tactic for getting more viewers is to flier the neighbourhood letting them know that the Street TV station will be retransmitting pay for view football matches for free – then show radical content at half time. Outrageous!

★ Squatted venues: Often squatted venues give you the flexibility to provide an ideal screening environment. People often feel a good sense of ownership over the space and this can help people really respond to social change films.

★ Cinemas: Sometimes independent cinemas let you use their facilities and you may be able to link something up with an existing festival.

Promote the gig tactically and add performance into the mix

Performance, live music and having a good time are a great way of getting the message across. Start up something new and try to get a buzz going. Get a popular local musician, DJ or VJ, poet or MC to come along. Make sure you know that there will be events in the lead-up to your screening to give out fliers. You will normally find a lot of people who will play for a good cause. Choose DJs and musicians to suit the mood of what you are doing. Partner up with other events; try to involve other promoters too.

Techniques to make screenings more stimulating, interactive and less formal

★ Stopping films and getting feedback
★ Offering choices of what film to see next
★ Having the films introduced by relevant people
★ Q and A sessions with people who where involved in the film
★ Plants in the audience to stimulate debate
★ Using extracts of video for 'what happened next' role plays.

If you pull together a diverse range of people to watch it, then discussions can keep going for as long as the film itself. It might be good to make sure that someone with facilitation skills is there.

Choosing films with an audience and activity in mind

It can be a good idea to think of films to communicate a message to a specific group. Specifically targeted fliers, mail shots to email lists, and personal contact and announcements at events work well. You can work with groups to chose a video with a suitable social change message. Hopefully the group you are working with will have suitable activities and projects to promote at the screening.

Programming films

There are lots of different films that can be used as part of a schedule of a screening night:

   ☆   Short action films
   ☆   Unedited rushes (works well when you have someone to narrate it)
   ☆   Sections of longer films
   ☆   Longer films, well crafted films.

How to source videos for your screenings/video archive

Often if you are doing screenings you may be asked to come up with 30 minutes of films on a particular issue – privatisation, climate change or squat actions. This is made a lot easier if you have a large archive of films. This may take the form of a library of DVDs, CDs and VHS tapes. It may also involve having a hard drive full of digital films which have been downloaded from the internet or networked hand to hand. There are lots of places where you can get material.

Offline content

   ☆   Traditionally the best way to get good content for screening is to order VHS tapes and DVDs from mail order catalogues, or even get film prints delivered and hire out your local cinema (see www.culture shop.org).

   ☆   Film screeners now often tour with a DVD or their work which they screen and sell. They also send out copies to screeners that they can't visit in person but who find out about their film through their touring.

   ☆   Hand to hand digital distribution has made it a lot quicker to network films suitable for activist screenings. It has become quite common for screeners to spend break times in activist gatherings frantically swapping digital video files and DVDs.

   ☆   'Ruff Cuts' and the 'European Newsreal' are two projects which provide a high quality source of shorter action based films. These projects have produced between them 40 or so CDs of radical video content. The films are all distributed under the 'Copy-left license', meaning they can be freely copied and passed on.

Online content   For a wealth of online content, see the resources at the end of this chapter.

Some tips for good distribution
You can also help to create a network of film screeners for your subject of interest by co-coordinating dates with other venues, doing joint publicity, previewing copies, paying independent film makers and getting material in sympathetic shops.

conclusions

Since the advent of the internet and a reduction in costs of audio-visual equipment, producing our own media has become considerably easier. All of the ideas mentioned in this chapter are united in promoting media work based on freedom, co-operation, mutual aid, justice and solidarity. From community and neighbourhood campaigns, direct actions, grassroots mobilisations to critical analysis of the world we live in, building alternative and independent media gives us a voice with which to shout. Producing and disseminating our ideas, inspirations, news and creativity has an enormous potential role to play in constructing and renewing a sense of community. This chapter has contained a few ideas to get you started and the resources section will help you put your ideas into reality.

Mick Fuzz is a Manchester-based media activist who has worked extensively with video and multimedia projects at community and international levels. He is co-founder of Beyond TV, a video documentary project, and Clearer Channel, a web-based video sharing project.

resources

Books
Chomsky, N. (1997). *Media Control. The Spectacular Achievements of Propaganda*. New York: Seven Stories Press.
Chomsky, N. (1998). *Manufacturing Consent: The Political Economy of the Mass Media*. New York: Pantheon Books.

Edwards, David and David Cromwell (2005). *Guardians of Power: The Myth of the Liberal Media*. London: Pluto Press.

Gregory, S. (2005). *Video for Change. A Guide for Advocacy and Activism*. London.

Harding, Thomas (2001). *The Video Activists Handbook*. London: Pluto Press.

Lewis, Jeff (2005). *Language Wars*. London: Pluto Press.

Lloyd, John (2003). *What the Media are Doing to Our Politics*. London: Constable.

McChesney, Robert W. (1997). *Corporate Media and the Threat to Democracy*. Open Media Pamphlet Series. New York: Seven Stories Press.

McChesney, Robert W., Russell Newman and Ben Scott (eds) (2005). *The Future of Media: Resistance and Reform in the 21st Century*. New York: Seven Stories Press.

Monbiot, George (2001). *An Activists Guide to Exploiting the Media*. London: Bookmark.

Nichols, John and Robert W. McChesney (2002). *Our Media, Not Theirs*. New York: Open Media.

Project Censored (ed.) (1999). The *Progressive Guide to Alternative Media and Activist*. Open Media Pamphlet Series 8. New York: Seven Stories Press.

Sakolsky, Ron and S. Dunifer (2001). *Seizing the Airwaves: A Free Radio Handbook*. Oakland, CA: AK Press.

Websites

Local newsletters
Bristle www.bristle.org.uk
Haringey Newsletter www.haringey.org.uk
Rough Music www.roughmusic.org.uk
The Pork Bolter www.eco-action.org/porkbolter
Walthamstow Underdog www.libcom.org/hosted/wag

Participatory video
Carbon Trade Watch www.carbontradewatch.org/
Insight www.insightshare.org/
Raised Voices www.raised-voices.org.uk/

Video/media resources
Beyond TV www.beyondtv.org
Clearer Channel http://clearerchannel.org
Creating Online Video http://en.wikibooks.org/wiki/Video
Culture Shop www.cultureshop.org
Digital Video Archive www.ngvision.org
Engage Media – Australia http://engagemedia.org
Eyes on IFI www.ifiwatchnet.org/eyes

Indymedia Translation Project http://translations.indymedia.org
Online News Archive www.chomskytorrents.org
Online Video Producers www.transmission.cc
Open Source Internet TV www.getdemocracy.com
Our Video Toolkit www.ourvideo.org
Undercurrents Video Collective www.undercurrents.org
Video Indymedia http://video.indymedia.org
Video Syndication Network http://v2v.cc
Video/Image/Text archive www.archive.org

Independent media/news

Alternative Press Index www.altpress.org
Global Indymedia Network www.indymedia.org
Indymedia Documentation Project docs.indymedia.org
LibCom www.libcom.org
Melbourne Based Community Radio Station www.3cr.org.au
Schnews Direct Action Newssheet www.schnews.org.uk
UK Based Spotlight on Corporate Distortions of Mainstream Media www.medialens.org
US Based Alternative Press Index www.altpress.org
US Based Centre for Media and Democracy www.prwatch.org
Znet www.zmag.org (Huge archive of radical news and analysis.)

Open source/publishing

Creative Commons Licenses www.creativecommons.org
Open Publishing Explained www.cat.org.au/maffew/cat/openpub.html

17 why we need to take direct action

Alice Cutler and Kim Bryan

Direct action is an important part of political activism. But its exact definition is quite elusive, meaning to act directly to address an issue of concern. It stands in contrast to indirect or political action where elected representatives are asked to provide a remedy on our behalf. A huge array of things in this book could be considered 'direct action' in the wide sense of the word; it can be a philosophy for life which impacts on the way that we organise our health, our education, or the way we organise and communicate. So what makes this chapter different? Here we turn to look at campaigning for change. Direct action has been a vital catalyst in struggles for social justice and, instead of making demands of the authorities, occurs when people place their bodies and their freedom in the way of power. It is when people do not wait for electoral change; when, for example, people refuse to sit down on a plane deporting an asylum seeker; people occupy and restore empty buildings in response to the lack of affordable housing and speculation; detainees go on hunger strike to demand release and an end to inhumane conditions; instead of marching against war, people occupy an oil company's headquarters; or people blockade a train carrying nuclear waste. Acting directly means not deferring your personal ability, power and responsibility to pre-existing structures, but doing it yourself.

It's important to note that direct action is an emotive and controversial topic and is difficult to clearly define. For example, since it is essentially just a tactic, it can be used by many different groups (on the political left and right, both extremist and pacifist) in a variety of ways (blockading, striking, organising) for a variety of political ends (workplace change, community empowerment, uprisings). This chapter, however, mainly concentrates on direct action as used by autonomous groups in the UK, who mainly organise horizontally, collectively and without leaders, who we wouldn't expect to want to seize state power nor use violence against individuals to achieve their aims. In this context, we look at why we need to take direct action, at some of the successes and potentials while also attempting to deal with some of the limitations and criticisms. The intention is not to define what is 'legitimate' and 'not legitimate' direct action, but to explore these debates and tactics as they are used in reality.

Box 17.1 Civil disobedience or direct action?

The two terms are often used interchangeably but there are some significant differences. Civil disobedience is the active refusal to obey certain laws, demands and commands of a government or an occupying power. It was significant in non-violent resistance movements in India in the fight against British colonialism, in South Africa in the fight against apartheid, in the American Civil Rights movement in the fight against segregation and disfranchisement and in the anti poll tax movement in the UK. Direct action is sometimes, but not always, a form of civil disobedience. Not all forms of direct action involve breaking a law and they are not necessarily open or public.

why take direct action?

If voting changed anything, they'd make it illegal. (Emma Goldman)

Historically, very few political movements have relied exclusively on legal means, most have employed a mixture of 'direct' and 'political' action to reach their goals. Reforms of law and governments can have huge effects but real systemic change and action has historically emerged from struggles from below. Many of the rights that citizens have in Western democracies now – for example the right of women to vote, the eight hour working day, the working age limit, the end of feudalism – were all forced by massive upheaval and resistance by ordinary people. The term 'propaganda of the deed' stems from nineteenth-century anarchists and libertarians which for them meant taking direct action to change the world and inspire others to act.

Choosing to act directly is often combined with the belief that campaigning for change through representative democracy is ultimately futile. Many anarchists and social change activists believe that whether an elected government is right wing or left wing they exist in a system which is made up of entrenched positions of power and

Box 17.2 *The struggle for women's suffrage*

In the UK, the Women's Social and Political Union (WSPU) or suffragettes, as they are more popularly known, fought for equality for women and the right to vote, under the slogan 'Deeds Not Words'. From their beginnings in 1903, they were a militant organisation and their actions included burning down churches, breaking windows of shops in Oxford Street, refusing to pay their taxes and, once, sailing up the River Thames hurling abuse at Parliament as it sat. On the face of it, the suffragette movement was a triumph for direct action as women finally achieved suffrage on the same terms as men in 1928. During World War I, women were urged into the factories so that men could be conscripted to fight. The suffragette movement split and the WSPU called a 'ceasefire to the campaign' for the duration of the war whilst the Women's Suffrage Federation, opposed to the war, continued the struggle. The political movement for women's suffrage began in earnest in 1919 and, once it became apparent that granting women the vote represented no threat to the underlying power structure, the reform was passed.

influence. An established order exists of landowners, legal systems, police powers, educational academies, the military and companies that control production and natural resources, which is founded on historical inequality. Lobbying governments for change is fundamentally limited therefore in what it can achieve as established power lies beyond democratic selection. In addition, many feel there are so few choices at the ballot box that people themselves must take control. The many examples of left-wing governments promising reform while in opposition and then failing to deliver when in power strengthens this rejection of parliamentary democracy. Short-term governments mean that policies to tackle long-term challenges, such as climate change where solutions are considered to be 'unpopular', further exacerbate notions of the democratic deficit. Power from above, from international institutions, such as the G8 and the World Trade Organisation, with national governments acting as willing partners, savagely promote global neoliberal market policies. This further reduces the likelihood of policy emerging which can deliver bold change. Examples of these policies which work against progressive change include Structural Adjustment Programmes which brought forced liberalisation and privatisation to developing

countries, and more recently policies such as Private Finance Initiatives (PFIs) in the UK where access to previously cherished national public assets, such as schools and hospitals, are liberalised through legislation set down by the WTO.

Box 17.3 Coalition in Defence of Water and Life, Bolivia

The city of Cochabamba in Bolivia has become a key symbol of a victory of people's struggle against global capitalism. A huge majority opposed the privatisation of their water supply by the US transnational Bechtel, which meant prices increased by up to 400 per cent. A broad-based movement of workers, peasants, farmers and others created La Coordinadora de Defensa del Agua y de la Vida (the Coalition in Defence of Water and Life, La Coordinadora for short) to 'de-privatize' the local water system. Hundreds of thousands of Bolivians marched to Cochabamba in a showdown with the government, and a general strike and transportation stoppage brought the city to a standstill. Despite a severe police response, finally, on 10 April 2000, the directors of Aguas del Tunari and Bechtel abandoned Bolivia, taking with them key personnel files, documents and computers, and leaving behind a broken company with substantial debts. Under popular pressure, the government revoked its hated water privatisation legislation. Deeply chagrined at the failure of its pet project, the local government basically handed over the running of the local water service, SEMAPA, to the protesters and La Coordinadora, complete with debts. The people accepted the challenge, and set out to elect a new board of directors for the water company and develop a new mandate based on a firm set of principles: The company must be efficient, free of corruption, fair to the workers, guided by a commitment to social justice (providing first for those without water), and it must act as a catalyst to further engage and organise the grassroots.

Source: Adapted from Barlow 2001, *Blue Gold.*

Direct action tactics can be used to various ends: to amplify voices of opposition, to prevent something from happening or to force a change in policy. There are many different types of action that have multiple aims. Some of the most high profile actions are carried out by large NGOs, such as Greenpeace, who use trained activists on a range of environmental campaigns from ships trying to stop whaling to disrupting trading on the London Petroleum Exchange. These are often highly mediated actions with a huge publicity budget and argue that 'If there's no photo it didn't happen.' Other groups are less interested in media coverage but aim to physically stop something from happening, for example, by sabotaging fox hunts or stopping anti-fascist rallies and marches. Others, such as covert acts of sabotage, are completely secret and are not publicised but aim to challenge specific targets, such as 'night time gardening' (digging up genetically modified crops).

The majority stand somewhere in between – banner drops and some blockades are more symbolic and although they may cause disruption are primarily a way to raise awareness of an issue and generate debate. Street parties aim to highlight how much more sustainable the world would be with fewer cars, but also bring people together

Box 17.4   Community organising: Residents Take Over!

'Community organising and residents groups often conjure up images of people moaning about dog-shit, broken street lights or fly-tipping – and what's that got to do with changing the world? Fair enough question to ask. Yet it is these small, community led discussions, exchanges and actions that are the seeds of building a new society, and spreading the ideas of mutual co-operation beyond a small clique. What I like about people involved in residents groups is, if you say to them: "we should be independent, build up community spirit, support each other and co-operate, we are all equals, we should make all the decisions about our area together, with the decisions based on our community's real needs", nearly everybody would agree – it's all common sense! In fact, such common sense ideas are actually a radical basis for alternative politics, for a real counter-power and a new society if acknowledged and built on. Through residents groups people are effectively able to directly challenge, influence and eventually make all the decisions which affect them and their communities, for example to resist anti-social development schemes or cuts in local services. A community is created that is strong, supportive and empowered and people are acting on a basis of co-operation and solidarity with each other.'

Source: Dave Morris, Haringey Solidarity Campaign www.haringey.org.uk.

and build a sense of community. Guerrilla gardening or 'green blocs' actions aim to impact positively on a local area by creating gardens on land left to decay. One of the lasting effects of any given direct action is its ability to inspire and encourage others to think and take action themselves.

While campaigns often use a range of tactics, including publicity, outreach and lobbying, direct action has often worked far more quickly and effectively than indirect action. The campaign against genetically modified (GM) food in the UK is one example where direct action was used as a catalyst for a wider social movement for change. Small groups of people who began pulling up crops, both openly and covertly, and trespassing on land contaminated with GM foods not only removed the source of contamination but also attracted much media coverage. In turn, this raised public awareness and a range of campaigning tactics generated enough public pressure to eventually force the UK government and EU to impose a five-year moratorium on GM crops in 1998.

Direct action is certainly not the preserve of anarchists, leftists or libertarians. Indeed, anarcho-syndicalists, who coined the term 'direct action' in the US labour struggles of the 1900s, considered that their bosses used direct action tactics, such as lockouts and cartels, against them and in response they turned to strikes and sabotage in the workplace. As the centre ground of politics becomes less tenable, groups right across the political spectrum are using direct action to show their power and to bring the state to the negotiating table. Recent examples include: a Countryside Alliance demonstration against the banning of fox hunting in 2004 in which 10,000 people clashed with police outside the UK Parliament while four members managed to breach security and enter the House of Commons; French lorry drivers bringing the country to a standstill in 1992 over new charges to their licences, while their English counterparts did the same over rising fuel prices, and Fathers 4 Justice, who climbed Buckingham Palace in London dressed as comic book superheroes to highlight their demands for access rights to their children.

movements for change

This chapter now looks at three areas where direct action has been used by mainly libertarian, horizontal political groups: the anti-roads movement, workplace action and anti-war protests. These examples are but a few of the huge collection of movements of resistance across the world. There are many other movements we could look at, such as the No Borders network, who oppose deportations and fight for freedom of movement, anti-capitalists, feminists, animal liberation, anti-racism and queer groups. While the ones we look at below are all different, the lessons from them are intertwined. These contributions are included in order to give a taste of the breadth and scope of resistance and have been written by people involved within these movements.

Figure 17.1
Action against Shell, 2006

Source: London Rising Tide.

The anti-roads movement

Margaret Thatcher's reign of power through the 1980s and early 1990s saw the introduction of a £23 billion programme of intensive road building. In the 1990s environmentalists began to organise and use direct action tactics to campaign against these proposed roads and raise awareness of car usage and pollution in Britain. The anti-roads movement protest camps were, in part, inspired by the Earth First! movement originating in the USA, a loose network whose general principles are non-hierarchical organisation and the use of direct action 'to confront, stop and eventually reverse the forces that are responsible for the destruction of the Earth and its inhabitants'. In 1991 the first UK road protest camp was set up in Twyford Down, Hampshire, against the proposed extension of the M3, while others followed in Newbury, East London, Stanworth Valley, Pollock Park and Birmingham, to name a few.

A protest camp is built to purposefully block a development before any construction can take place. Camps serve as a launch pad for actions, from digger diving to occupying offices of contractors. People living on camps provide a constant visual reminder of the protest and the issues at stake, inspiring people to take action and pride in their area.

The labels 'eco warriors' and 'tree huggers' emerged as people took part in camps and protests against the building of roads and runways across the UK, setting into action a chain of events which heavily influenced government policy and marked public opinion. Television images of people being dragged from trees helped spread the idea and camps began appearing in countries, such as Poland, Ireland and the Netherlands.

> The seeds were planted for a movement that inspired and mobilised thousands of people to take direct action in defence of mother earth who was being raped and killed by corporations and governments in favour of so-called progress. (Mik, road protest campaigner, interview with authors)

As the 1990s drew to a close a change of government and the success of the anti-roads movement meant that the road building programme was temporarily halted and 110 projects were shelved. Many of those involved with 'eco-action' wanted to deal with what they perceived as the root of the problem. In the late 1990s anti-capitalist protests began to overshadow ecological direct action as people increasingly equated road building and unsustainable developments with the wider economic system. One group which emerged from the anti-roads movement was 'Reclaim the Streets', which organised street parties and attempted to combat car culture in the cities. The largest of these was the Carnival Against Capitalism in the City of London on 18 June 1999, which an estimated 5000 people took part in and brought the City to a standstill. Countless other events took place simultaneously in 43 countries. Protest camps remain a key and crucial form of resistance to proposed developments – today there are numerous ecological and peace protest camps in existence around the issues of roads, supermarket expansion, gas pipeline developments and nuclear submarine bases. The lessons and experiences from the anti-roads movement have been an inspiration for many people to join together to resist unsustainable developments in their areas.

Direct action in the workplace

As long as work and wage labour have existed there has been direct action in the workplace. Around the world, industrial developments are followed by a wave of struggle by workers, forcing wages up and capital's search for profits on to new terrains. For example, in the 1920s the car industry's first boom in Flint, Michigan, USA, was followed by strikes and occupations which brought the whole production process to a standstill. In 2006, in the booming industrial areas near Delhi, India, strikes at Honda factories and suppliers pushed up contract workers' wages by nearly 50 per cent in one year.

The workplace is where profit is extracted from workers, but it is also where these profits are vulnerable. Direct action in the workplace is not only at the forefront of a struggle for a better life, but it is a major site of struggle for an end to the entire economic system. Previously, the huge Fordist factories in the 1960s and 1970s were strongholds for co-operation and mobilisation amongst workers. The more complex supply chains and just-in-time production which have emerged since the Fordist era were attempts to dismantle this, but worker co-operation and struggle continues. For example, the Liverpool dockers' strike between 1995 and 1998 not only saw massive local support, but attracted international solidarity from other longshoreman unions and days of action around the world.

Workers can self-organise, challenging the role of the boss and the worker. The occupied factories in Argentina are an inspiring example. Leading up to and after the 2001 crisis, factories and businesses were closing by the dozen leaving millions unemployed. Used to being at the rough end of the global economy, many Argentinian workers have reclaimed factories and businesses ranging from five-star hotels and pastry factories to metal works. Each factory or business has its own story, but what they have in common is that most organise using assemblies and flat structures to make decisions. In a country where 60 per cent of people live below the poverty line, reclaiming a factory or business, running it autonomously and horizontally, and everyone getting fair wages is an empowering and life improving activity. Many of the reoccupied factories are co-ordinated through two organisations: the National Movement of Recovered Factories, which has 3600 workers across 60 factories, and the National Federation of Workers' Cooperatives in Recovered Factories, which has 1447 workers across 14 factories. Some of the most inspiring examples include Hotel Bauen, a former five-star hotel in central Buenos Aires, the Brukman Textile Factory and Zanon Ceramic in the Neuquén province.

Collective direct action in the workplace is different from appealing to any of the various mediating bodies: the unions, works councils, left-wing political parties and NGOs. Whilst they maintain the illusion of democracy, their primary interest is in their own image and survival, and they are limited to legal struggle. The law, far from being neutral, has been carefully crafted to weaken workers' collective power – for

Figure 17.2   Workers' assembly at the Zanon occupied factory, Argentina

Source: Paul Chatterton.

example, the laws against secondary picketing in the UK to avoid actions spreading or the recent US Homelands Security Act banning strikes in the public sector under the guise of anti-terrorism laws. Taking direct action is often much more effective and empowering. Wildcat or unofficial strikes, when workers decide to strike without the sanction of a labour union, either because the union refuses to endorse such a tactic or because the workers concerned are not unionised, are unpredictable and can force bosses to meet demands. The Hillingdon workers went on strike between 1995 and 2000, holding out against their ex-employers not to mention their own union in the struggle to win back their recently privatised hospital jobs at the same rates of pay. The strikers picketed day in and day out, enduring wind, rain and too many torrents of abuse, and it eventually became the longest strike in British history.

Temping agencies and outsourcing have come to dominate the labour market and can divide and confuse us, and make us precarious and vulnerable as we sell our labour. However, janitors in the USA worked collectively to demand wage increases although they officially worked for different agencies. In taking collective direct action in our workplaces we can not only improve our daily lives, but also learn our strengths, how to organise and co-operate together and so catch a glimpse of how life without capitalism might be.

The anti-war movement

It won't be union leaders or paper sellers or 'organisers' that will stop this war. It will be ordinary, angry active people – us, you, your neighbours, and your mates – taking direct action. Stopping high streets at rush hour. Shutting down government and military buildings. Having sit down protests on marches instead of moving on whenever the police tell us to. (*Schnews*, 27 September 2002, Issue 3)

After months of marches, protests and mounting despair at the prospect, when the UK–US led coalition bombing of Iraq began, thousands of people around the world took direct action. The growing sense of frustration against the war led many people to a more radical analysis that it was being fought over oil and that the blood of innocent people was being spilt in order to keep the oil companies and US administration happy. And in exceptional times ordinary people do extraordinary things.

Ranging from thousands going on strike, children refusing to attend lessons and walking out of school to disarming war machines, a huge array of actions were attempted to stop the war and express their anger, to assert that the war was 'Not in My Name'. Whilst the protests didn't stop the war, they played an important role, as one student reflected: 'We didn't stop the war but an entire generation saw through the lies of the government and the media' (School Students Against the War 2006).

In Brighton, UK, in 2003, on the day the bombing of Iraq began, 5000 people took part in a mass act of disobedience on the main shopping street. The town hall was occupied and people refused orders to move by the police. One offshoot from Brighton anti-war actions is the campaign against arms manufacturer EDO-MBM. Activists' research revealed that the company were making components of the Pathway system used in the Coalition bombing of Baghdad, 'Operation Shock and Awe', and also in the Occupied Territories of Palestine. The Smash EDO campaign in Brighton has persistently targeted the factory that makes components for missiles with regular noise demonstrations, blockades, weapons inspections and rooftop occupations as well as many street stalls, information nights and other campaign outreach. They have also been battling against the criminalisation of protest. In March 2006 they won a high court case which cleared them of charges of harassment. Their defence was that EDO-MBM had been ancillary to war crimes. Their victory exposed collusion between the local police and the company, and showed the way that the protesters had been targeted with ridiculous allegations, 'because they dared to stand up and tell the truth about the arms industry and about the war crimes these arms dealers are involved in'. Their victory was significant as they were the first of 20 such cases to win and have the injunction overturned.

Many of those involved in the campaign have visited Palestine as part of the International Solidarity Movement (ISM). ISM is a Palestinian-led movement committed to resisting the Israeli occupation of Palestinian land using non-violent, direct action methods and principles. ISM aims to support and strengthen the popular resistance by providing the Palestinians with international protection and a voice with which to non-violently resist an overwhelming military occupation force. Witnessing the impacts of these missiles and having direct contact with the people affected strengthened the activists' conviction to campaign against corporations at home who profit from the war.

Figure 17.3   Demonstration in Billin, Palestine, against the building of the apartheid wall

Source: Keren Manor, Active Stills.

It is often both morally and legally justifiable to break a law in order to prevent a worse crime happening. Therefore whilst the tactics employed by anti war and peace movements may not always be legal, they see them as legitimate. In February 2003 during an action by Ploughshares, five people broke into Shannon Airport and dismantled a fighter plane destined for Iraq. After three years of legal wrangling and pressure from both the US and Irish Governments to prosecute, the five were eventually found not guilty and acquitted of all ten charges against them. The Ploughshares movement is made up of people committed to peace and disarmament and who non-violently, safely, openly and accountably disable a war machine or system so that it can no longer harm people. In August 2006, there was a public trespass at Prestwick Airport in Scotland by anti-war and peace activists carrying out citizen weapons inspections, angry at the use of the airport to transport US 'bunker buster' bombs to be used by Israel in its invasion of Lebanon. As a direct result of this action all further such flights were directed to military bases elsewhere in England, while activists in Derry, Northern Ireland, smashed up computers at Raytheon, manufacturers of Patriot, Tomahawk, Cruise and Sidewinder missiles.

Imagine if the millions of people who marched against the war around the world had taken direct action. If people had blockaded arms factories, air bases, refused to transport weapons when they learned that their cargo included bombs to be used in Iraq, as two UK train drivers did recently, or conscientiously objected to serve in Iraq on the grounds that the war is illegal and immoral. How would warring governments manage that wave? It would engulf their political processes and present a huge crisis of governance. Would there still be war now?

challenges to action

Measuring effectiveness

Measuring and quantifying the outcomes of direct action is extremely hard. What initially appears a great success can soon fade away as a one-off phenomena, what starts slowly and unsurely can go on to build and achieve great things. There is no arbitrator on which tactics will work. However, it is useful to evaluate the successes and failures of actions and campaigns at short, medium and long-term intervals as campaigns work on many different levels. For example, one critique of actions by anti-capitalists that aim at dismantling institutions, such as the G8, World Trade Organisation, International Monetary Fund and World Bank, are that they are unrealistic and unachievable. This may overlook smaller intermediate steps which activists take towards achieving larger goals. For example, mass street protests to shut down summits actually did work temporarily (for example, in Seattle in 1999, against the WTO, and Prague in 2000, against the IMF and the World Bank, where summits were seriously disrupted). Many of these global institutions are now struggling to regain their legitimacy in the face of sustained protests from global civil society. It is often the sustained pressure of actions, of repeatedly coming back, of defying, agitating and expanding that creates successes. These mobilisations brought people together, solidified networks of solidarity, allowed people the opportunity to exchange experiences and ideas, and join together on actions. It has led to the forming of what has become known as a global 'movement of movements' that is historically unprecedented – a loose network of groups working towards global justice which avoids the dogmas and hierarchies of Marxist-Leninist groups.

When is direct action justified?

It is important to talk abut the morality and, therefore, legitimacy of direct action. Suicide bombings, kidnappings and firebombing are all forms of direct action, and boycotts, sit-ins, occupations and community organising can be as well. There are a range of ethical positions that people are prepared to fight for and discussions on direct action tactics often end up being philosophical discussions on moral issues – some may be driven to act by passion and outrage towards what they see as morally defensible actions, whilst others try to build up a wider social mandate to avoid accusations of extremism or vigilantism. It's clear that certain forms of direct action involve different forms of violence, to yourself and to others. Some feel that violence, in some cases, is legitimate considering the violence we face everyday from an uncaring economic and political system. However, it's clear that direct action which in any form may directly lead to the loss of life is not defensible.

In the book *Desire for Change* (PGA Women 2001), the authors state: 'Non-violence has to be understood as a guiding principle, relative to the particular political and cultural situation. Actions which are perfectly legitimate in one context can be unnecessarily violent (contributing to brutal social relations) in another.' They go on to suggest that non-violence has very different meanings in India (where it means respect for life) and in the West (where it means also respect for private property). In many liberal democracies, public displays of confrontation are often seized upon by critics and the media as a way to depict all activists as mindlessly violent. This is often seen as a device to draw attention away from the issue at hand. To choose to be specifically non-violent could be understood to not allow for a diversity of tactics or even contribute to the criminalisation of part of the movement. For example, it would imply rejecting huge parts of the history of resistance in Latin America. The violence/non-violence debate continues to divide people. The question for any campaign is to decide, through discussion, on the strategic value of any action.

Repression

Governments are all too keenly aware that when people start acting directly for social justice and against all that oppose it, the legitimacy of power structures themselves are under attack. While the direct activist group's greatest strength is its adaptability, authorities learn about tactics and how to respond to them. Evolving movements are crucial, as political situations change, strengths and weaknesses come and go, the state brings in new laws and authorities learn new tactics. Tactics need to be constantly reviewed and adapted to different situations.

Since 9/11 and the war on terror, there has been a widespread crackdown on civil liberties. The hard won right to a fair trial, free speech and independence of the judiciary, and the global ban on torture are being challenged by the pronounced changes to the 'rules of the game'. These changes would make those who acted in solidarity with historical figures, such as Nelson Mandela or the suffragettes, culpable for glorifying terror. A UK example includes the Serious Organised Crime and Police Act which makes the entire square mile around Westminster, and anywhere else that the government chooses, to be a protest-free zone.

These are just some of the challenges that campaigning and direct action throw up. There are also considerations about the impacts on individuals and the emotional health impacts, which are looked at in the next chapter. The answers and responses to them depend on the framework and principles of individuals, groups and networks involved. Formulating responses and finding answers to some of the challenges can be achieved by evaluating, debating and discussing, and constantly looking for new and innovative ways to make ideas more accessible.

making ripples, creating waves

So where does this leave us? Sceptics may ask what's the point in shutting down one small factory or stopping a road being built? The company could just move elsewhere, the road can be relocated. What difference does it make if you shut down a petrol station to protest against climate change? It will be open tomorrow and only a few people may hear about it. But it is by joining together and resisting that a community finds a voice, informs people of its existence and proves that it is possible to stand up and challenge injustice. It's impossible to gauge the size of the wave you create from an action and building resistance – a wave could gather momentum, cross seas and borders, ripple down through generations and across oceans. Actions have inspired many and in their own small ways contributed to the building of more tolerant and just societies. Taking action involves stepping on the side of risk, spontaneity, possibility and creativity, and requires being willing to observe, explore and experiment. It is important to recognise the potential risks to personal liberty and safety, but history teaches us that the ability to create change exists when ordinary people stand up and act together for their beliefs. When people come together they start feeling that there are ways of changing the world and supporting each other based on principles different from profit and power. Whether it's about challenging the decision to knock down the public swimming pool to build luxury flats, occupying land to grow vegetables, blockade a summit of global leaders or going on strike against unfair treatment at work, taking direct action asserts a right to challenge inequality, corruption and unjust laws, and takes us a step closer to having control over our lives.

Alice Cutler and Kim Bryan are active campaigners who have been involved in direct action and campaigning around issues such as genetically modified foods, climate change, anti road and runway building, and building anti-capitalist networks and migrant solidarity. Other contributors to the chapter include Lady Stardust based in the Escanda community in Spain who is a political activist and writer, writing for a number of publications, such as Prol position *and* Wildcat, *and Mark Brown who works with London Rising Tide, a campaign group focusing on taking action on the root causes of climate change. Other information has been gratefully received from Haringey Solidarity Group, London Rising Tide, and the Brighton-based Smash EDO Campaign in the UK. This chapter draws upon the actions of thousands and many reflections and conversations from people who are too many and unknown to name here.*

18 how to build active campaigns

Kim Bryan and Paul Chatterton

This chapter is focused on building actions and campaigns that are about empowerment, self-management and mutual aid – doing it ourselves. Whilst most campaigns also depend a great deal on spontaneity, creativity and large amounts of luck, this guide is designed to explore some of the key aspects to consider in creating and running an effective campaign. The material for this chapter has come from a variety of sources and draws on aspects of different campaigns which have built effective resistance.

why are you passionate about it?

Outrage, indignation, 'fire in your belly', a desire to see change are just some of the reasons that people feel motivated to start a campaign. The outbreak of unjust wars, a government introducing an unpopular policy, a company quarrying a local park, a supermarket building on a street full of local shops, school children not having a safe place to cross the road, unfair working conditions?

Defining your aims and objectives is vitally important for communicating and being sure about your campaign and, depending on the issues, there will different points to consider in developing them. The Ruckus Society (2003) suggests a few options:

☆ To announce: To bring to light some scandal or shocking event.
☆ To reinforce: People can be aware of something, reinforcing is reminding them of something that they know already.
☆ To punctuate: There might be an event that needs remembering or to remind people that an issue has not gone away.
☆ To escalate: There might be a need to raise the stakes on an issue that is getting more pressing.
☆ To increase morale: A group might be ebbing low and needs a boast through taking action.

Box 18.1   100 ways to resist

Refusal of assembly to disperse. Sit down. Bodily interjections. Bodily obstruction. Trespass. Airborne invasion. Occupations. Inviting arrest/imprisonment. Sit-in. Stand-in. Ride-in. Pray-in. Return of waste products. Heckling. Guerrilla theatre. Protest strip. Graffiti. Subvertising. Refusal to collaborate. Declining government awards/appointments. Boycott of elections. Hunger strike. Ghosting. Publicising individual's activities. Social boycott. Ostracism. Denial of social/sexual relations. Excommunication. Boycott of meetings, events or lectures. Group silence. Walk-out. Picketing. Breaking social taboo. Harbouring fugitives. Sanctuary. People's public hearings and courts. Consumers' boycott. Withholding of rent. Refusal to pay tax. Refusal to pay debts or charges. Withdrawal of bank deposits. Blacking of goods by suppliers. Blacking of raw materials by workers. Demonstration strike. Go-slow. Work-to-rule. Bumper strike. Wildcat or lightning strike. Lock-up or stay-in strike. Reverse strike. Personal strike. General strike. Overloading facilities or services. Overloading administrative systems. Stalling by customers. Breaking bad laws on principle. Publishing secret material. Disclosing secret identities. Tracking. Forgery of letters. Breaking official blockades. Refusal to recognise appointed officials. Non-co-operation with police. Removal of street signs, door numbers. Closure of roads. Infiltration of institutions with spies. Electronic picketing. Spoiling or contamination of goods. Monkey-wrenching. Liberating animals. Failure to pass on information. Deliberate inefficiency. Industrial sabotage. Non-retaliation. Entryism. Alternative radio/newspapers. Alternative school. Selective patronage. Alternative economies. Selective refusal of entry. Alternative community with independent sovereign government.

what are your aims?

It is important to have a target audience in mind and design the messages, slogans, publicity and campaign approach accordingly. In any action or campaign it is important to focus on a few concrete demands and think about how achievable they are. Some groups start small aiming for victories which give confidence, others start with bigger objectives and less chance of victory but are more radical and far-reaching in their aims. If you're going to start something new, concentrate on positive things, things that can build up people's strength and limit the risk of demoralising people. Once there is trust in the group you can try and tackle the difficult stuff that takes a long time to make progress on. Working out how best to achieve your aims can be a real challenge.

Box 18.2 Comparing aims

Trident Ploughshares is a campaign to disarm the UK Trident nuclear weapons system in a non-violent, open, peaceful and fully accountable manner by taking direct action against installations and equipment involved in the Trident system. 'By doing so we aim to inflict significant damage and disruption on these installations and when arrested we take full responsibility for our actions. Our defence in the courts is generally based on the primacy of international law. We do what we can to publicise our actions and the response of the authorities so that public awareness of the UK's indefensible nuclear weapons policy is increased and more and more people either become disarmers themselves or actively support the movement in a whole variety of ways' (Trident Ploughshares, www.tridentploughshares.org, May 2006).

If working on a community campaign your aims could be to develop:

* strong and vibrant local communities throughout every corner of our towns;
* local solidarity and mutual aid, increasing residents' self-activity and their interest and involvement in community affairs;
* a safe, pleasant and green local environment including every residential street;
* a wide range of publicly accountable community facilities;
* decent and affordable housing for all;
* a situation where people are effectively able to challenge, influence and eventually make all the decisions which affect them and their communities.

how are you going to resource your campaign?

Making a campaign happen or taking action doesn't cost the earth. But you do need some resources and skills. Here are some useful things to take into consideration:

★ Fund raising: You're going to need some money for lots of things – leaflets, photocopying, travel, petrol, food, banner making material. One way to make some money quickly is a benefit gig. Book a cheap community centre, a social centre or back room of a pub. Find a band or DJ to play some music, sort out food and drinks, and charge a few pounds or ask for donations on the door (see the resources section in this chapter for further ideas).

★ People: An effective campaign needs people who are committed to work together and show solidarity. Never underestimate your collective potential.

★ Publicity: This is essential and there's lots of creative ways to do it. It could be your key resource.

★ Press contacts: If you decide to use the media, try and build up trust with sympathetic journalists and send them regular information. Make a list for press releases. Don't bombard people though, and make sure what you send is clear and well written.

knowing your issue

You need to do some research to give yourself and others a greater understanding of the issues at stake. There are many ways to do this, through websites, campaign groups, holding information meetings or talking to other groups. You can present this information in a variety of material including press releases, leaflets, websites, dossiers, newsletters, radio shows and emails, and don't forget to have the right information when talking to people on the streets or at meetings.

Find out about the legal implications of what you plan, although it's important not to let this distract you from focusing on the real issues. Even if something is 'illegal' on paper, it still may well be the best and most efficient course of action to win your demands. For example, blocking a road is illegal, but at the end of the day, it might secure a victory and only mean small fines or small and temporary legal charges. Keep a record of any conflicts or disputes to use later.

This checklist of questions may help when planning:

★ When is the best date for the action?

★ Are there any visits or events worth coinciding with or avoiding?

* ☆ What time of day would be best?
* ☆ Were there previous actions and what happened?
* ☆ What about public opinion – how can you tie your message in or challenge assumptions?

Box 18.3   *Turning the Tide*

Turning the Tide, who provide campaigning support to activists and campaigners, suggest the following considerations:

* * Communication: What information or messages do you need to communicate and to whom? (e.g. if working for refugees, the main message you might want to convey is 'these are people just like us'.) What means of communication do you have?

* * Persuasion: What would persuade your opponents to make the change you want? Who do they listen to and take seriously? What methods could you use to communicate to them directly or indirectly?

* * Coercion: If your opponents are not persuadable, what kind of pressures, non-co-operation or action would make them decide to change? How would you achieve this? Is there a need for direct action?

getting people involved and organised

There are countless ways of getting people involved. It is best to start small and build up. Arrange a meeting in your home or in the pub with a few people and bounce some ideas around. The key is achieving the balance between having time for people to explore and discuss the issues and not losing people by sitting around rather than getting on with things. Limit meetings to two hours and bring along food and drink. Box 18.4 (*overleaf*) gives some examples.

so what's the plan?

There is almost a limitless range of actions and campaigns that you can get involved in. Think creatively but do learn from what's gone

Box 18.4   *Getting organised and involving people*

Getting organised

* Clearly define your aims and objectives
* Do you need a bank account, web page, postal address?
* Network at events, by phone, pledges of support, email lists, set up exchanges
* After meetings send out minutes with what was decided, who took on doing what and the dates of the next action/meeting
* Organise training such as Seeds for Change's 'Planning a campaign: aims, strategy and tactics'
* Keep records of money, press, decisions, etc.
* Allocate roles/make working groups

Ways to get people aware/involved

* Film screenings/concerts
* Posters on notice boards
* Fly posting/stickers
* Giving out leaflets
* Banners in the street, out of windows
* Community events
* Benefit gigs
* Public meetings
* Educational outreach workshops/talks
* Websites
* Publicity stunts
* Street stalls

before. Are there groups to link up with or who have made their resources available on the internet or as guides? You've basically got lots of choices to make. Here are some issues you might want to consider:

☆ Do you plan a sustained or a one-off campaign?
☆ Are actions most effective if they are open or covert, large or small, in one place or many?
☆ Do you want the press there?
☆ Who in the group is willing to risk arrest? What are your limits and guidelines?

Here are some examples of actions that may be useful to your campaign.

Banner drops

Hanging a banner can be a great way of getting a message across to a huge range of people. It's a good way of getting people together to organise a banner painting session in your garage, community or school art space. You'll need lots of bed sheets or material. Try simple, large slogans like: 'Support Our Troops. Bring them Home' or 'No car park on the moor'. Try and make it funny or connect it with local issues. Brainstorm short and simple slogans, which will be inspiring to a wide audience, and sketch them out before you paint. Always use water-based paints. Make good holes in each corner and reinforce with eyelets. Cut holes throughout so the wind can pass through. Pick a good location where lots of people will see them and where it's also difficult for the police or curious locals to remove. The best time to hang the banners is early morning so you catch the morning rush hour. Once the banner is up take photos, write a report, post on Indymedia and send a press release.

Direct action in the workplace

Strike!   A timed and strategic strike, e.g. 'sit-downs' when everyone just stops work or everybody leaves work to go to the boss' office to discuss some matter of importance can be best. For longer strikes to be successful it is vital that the work is not simply done elsewhere. Blockading the workplace, leafleting scab labour temp agencies, making links with other sites of the same work so that they know what is happening when they are asked to sort mail and taking phone calls are vital to the success of the strike. Everyone calling in sick on the same day can also be used.

Working to rule   Following your instructions to the letter and nothing extra can result in the workplace becoming chaotic and it is hard for management to discipline you for this. Doing your job, just very slowly and carefully. You can also take direct action by not using equipment if it is unsafe, not wearing name badges, not sitting in newly allocated seating arrangements, and taking the breaks allotted to you in your contract even if the boss says you can't.

Good work strikes   Providing better or cheaper service at the boss's expense can make your struggle popular. For example, workers at Mercy Hospital in France refused to file the billing slips for drugs, laboratory tests, treatments and therapy. The patients got better care, the hospital's income was cut in half and panic-stricken administrators gave in to all of the workers' demands after three days (see http://libcom.org/organise/goods-work-strike). In 1968, Lisbon bus and train workers gave free rides to all passengers to protest a denial of wage increases. In New York City, restaurant workers, after losing a strike, won some of their demands by 'piling up the plates, giving them double helpings, and figuring the checks on the low side'.

Pie-ing

> I am used to the custard pies, I am even beginning to like the taste of them. (Bono, *Independent*, 16 May 2006)

Putting a big fresh pie in the face of someone responsible for something that has annoyed a lot of people is an easy, effective and often hilarious way to take direct action. Aim for someone who's rich and famous and culpable – Clare Short for her policies on international development, Mark Moody Stuart for his tribulations as ex-head of Shell, and Milton Friedman for championing free market economics. Choose a good, wet pie with lots of fruit and cream in it. Dress the part especially if you are to do it in a hotel. Your pie can be concealed in a brown paper bag or briefcase. Blend into the crowd and adopt a pleasant, civilised demeanour. It's difficult to both throw a pie and document the action so get an extra pair of hands to grab a photograph or video footage. Some pie assassins work in teams, others prefer to work solo. As you hoist the creamy pastry into the face of your prey, a quick quip can highlight your action. Afterwards, expect shock and chaos and get out of there quickly. Realise that you may be arrested or detained.

Blockading

Blockading means to either put your body or an object in the way to prevent something or someone entering or leaving. From farmers driving tractors into position to determined people locking on in the streets, it's always been a popular choice and there's lots of ways to do it:

★ Arm linking: You won't last long but if there are a few of you it can take a while. Once you have been dragged off remember that by totally relaxing and going floppy you become a dead weight and more people will be needed to carry you.

★ Arm tubes: Tubes made from plastic or metal piping, the diameter of a clothed arm, are a versatile tool. They need to be the length of two arms, ideally with a strong metal pin welded in the middle. You need to link your arms together inside the tubes, either with handcuffs, loops of strong cord or climbing tape with karabiners encircling the object. Be aware that if you lock on with handcuffs, you won't be able to release yourself. Arm tubes have been used to blockade gateways, roads and even airport runways. To remove you, the tube must be cut using hacksaws or angle grinders.

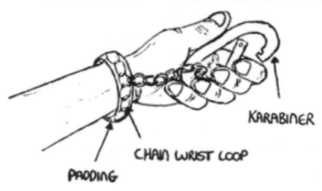

Figure 18.1   Karabiner hand lock on

Source: Road Raging for Beginners.

★ Bicycle D-locks: They fit neatly around pieces of machinery, gates and your neck. It is worth working in pairs when trying to lock on. If locking on to a machine, someone must let the driver of it know. It is important that anything you lock on to cannot be removed or unscrewed.

★ Handcuffs: These are particularly good underneath machines if you can find inaccessible bits to lock yourself to. Loops of strong cord or tape can often be just as effective and are cheaper.

★ Tripods: These have successfully been used as a mobile, easily erected blockade. They are made from easily obtainable materials, such as scaffold poles from building sites or long, straight tree trunks.

Figure 18.2   Tripod

Source: Paul Chatterton.

Street party

1. Get together with some like-minded people. Work on a plan of action. Sort out different roles, jobs and time scales. Imagine. What's possible?

2. Decide on a date. Give yourselves enough time. Not too much – a 'deadline' is a great motivator – but enough to sort out the practical issues, such as materials, construction, etc. You may need money.

3. Choose the location. Your street, the town centre, a busy road or roundabout, a motorway! A separate meeting place is good – people like mystery and it works as a decoy.

4. Publicise! Use word of mouth, leaflets, posters, email, carrier pigeon. Make sure everyone knows where and when to meet. Posters and paste go well on walls, billboards and phone boxes. Leaflet shops, clubs, pubs – everyone, and your mum.

5. Sort out your sound system. A party needs music – bike powered sound systems, rave, plugged-in, acoustic, yodelling – go for diversity. Invite jugglers and clowns, poets, prophets and performers of all kinds. Ask campaign groups to come along and set up a stall in the middle of the road.

6. How will you transform the space? Huge banners with a message of your choice, colourful murals, bouncy castle, a ton of sand, paddling pool for the kids, carpets, armchairs. The materials and money from earlier may come in useful here. Print up an explanation for this 'collective daydream' to give to participants and passers-by on the day.

7. For opening the street – or rather stopping it being reclosed by the traffic – ribbons and scissors are not enough. A large scaffold tripod structure with a person suspended from the top has been found to be useful. Practice in your local park. Block the road with a car or other barricade that can then be dismantled.

8. Have a street party! Enjoy the clean air and colourful surroundings, the conversation and the community. Bring out the free food, dance, laugh and set off the fire hydrants. Some boys in blue may get irate. Calm them down with clear instructions.

using the media

Whether you communicate with the media depends on the type of action you are doing. Although you can't control what they write or broadcast, it can help to get a message across, raise awareness and encourage more people to get involved. Talking to the media can be fairly scary. Help build people's confidence in talking to the media, create lists of FAQs and role play the answers, watch and review other people's interviews – all are good ways of ongoing media training. There are some great resources about how to deal with journalists and give good interviews that can be found online, such as the Media Toolbox, which outlines how to write press releases and deal with interviews (see www.gdrc.org/ngo/media/index.html).

organising on the day

Don't forget that if you are planning a day of action then you need to make sure you do some preparations to make it run smoothly. Think of key aspects, such as:

★ Legal support: If you are doing actions which will infringe certain laws then you need to be prepared – inform yourself, inform others of the risks, be prepared to deal with arrests and the police, sort out support for those arrested.

★ Stewards: You might want to have people who can identify themselves to help with traffic, media and police liaison, and any stressful or emergency situations.

★ Food and water: Make sure people are well fed and don't become dehydrated.

★ Communication: If there's a large group it's worth having some walkie talkies or mobile phones. Other techniques can be used like identifying group leaders with flags or flowers, or having runners between delegates.

★ Transport: How will people get there? There may be a few of you going to a remote location.

campaigning problems

There are a number of common problems to watch out for in sustained campaigns that can cause tensions and rifts in groups. Active campaigning, as well as being rewarding, can at times be very hard.

There are inevitably hierarchies of experience within horizontal projects working towards social change, which can lead to newcomers feeling inadequate and hence reluctant to get involved. These hierarchies develop totally unintentionally but can lead to imbalance and bitterness within a group. It is important that any group working horizontally spends time working on the personal dynamics of the group and that macho or controlling behaviour is confronted in a positive way that is ultimately a learning and beneficial experience for the group. Campaigning can be physically and emotionally demanding and fears about the implications of breaking the law can create further

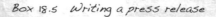

Box 18.5   Writing a press release

Good press releases are a useful tool in a campaign — remember
that news desks receive hundreds a day so they have to grab their
attention.

* Mark NEWS RELEASE clearly at the top — add your campaign name,
  phone number and logo.
* Next, write the date of issue and mark FOR IMMEDIATE RELEASE
  unless it is embargoed until the date of the action. An embargo is a
  note at the top of the press release telling journalists not to leak
  or print the story before a particular deadline. However, never trust
  the press to keep them. When publicising an event, make sure the
  press release is out well in advance.
* Use a snappy headline — this should capture the reader's attention.
  Limit to eight words.
* Some ideas for creating catchy headlines:
  (a) Use alliterations (Analytical Anarchists Answer Ahern)
  (b) Use colons (Mayday: More like a Circus then a Riot)
  (c) Offer tips (101 Ways to Avoid the Worst Effects of Climate
       Change)
* Include a summary in the first paragraph, including WHAT is happening,
  WHERE, WHY, WHEN and by WHOM. It needs to immediately grab an
  editor's attention or will be binned.
* The press release should be short, factual and well written. Avoid
  opinionated rants and jargon.
* Use short paragraphs and simple sentences. Keep to one, or two at
  most, pages.
* Use a quote by an identified person to tell your side of the story.
  Use pseudonyms if you do not want your name in the paper.
* Write ENDS at the foot of the press release. Ensure that there is a
  reliable contact with phone number on the release. This could include
  on site mobile phone numbers. If you want the contact details printed
  in newspapers it must be in the main body of the text.
* If your press release is for an event, press conference or photo
  opportunity, include a map or directions.
* If you do not want to go into massive detail on an issue in the main
  body of the text, but think it is of interest, include a Notes to
  Editors section at the end of the press release.
* Follow the press release up with a phone call to make sure that it
  was received.

barriers to taking action. Training, skills sharing and information nights are excellent ways to overcome these difficulties so people can develop confidence together.

The adrenalin of actions can also be addictive and many come to feel they are indispensable fearing that if they stop the whole campaign, camp or mobilisation could fall apart. In this context, burn out through overwork is a serious mental and physical problem. In many cases, people have had to withdraw for several years to recover from stress as a result of not looking after themselves. The sort of tunnel vision required to fight a campaign over time can leave people out of touch with 'normality'. Specialisation can mean that engagement with anything not directly related to the cause is avoided and objectivity is lost. Increasingly, activists are taking emotional issues a lot more seriously and are making sure that actions are backed up with solid support – both emotionally and practically.

The issues mentioned above all affect the health of a campaign. Make sure you have good support, clear aims, openness, regular skill sharing events, clearly defined roles, transparent finances, and time to have fun and relax with each other as well as do all the serious stuff.

evaluation and what next?

It's important not to just let a campaign fizzle out or lose momentum after an action. Holding an evaluation or debriefing is a good way to learn from the experience, recognise successes and weaknesses, and move on together. Decide on the scope and level of evaluation. It's often a good idea to get someone who is sympathetic but was not involved to facilitate.

Turning things around does not necessarily come quickly or immediately. The techniques of organising campaigns used in this chapter have been taken from a myriad of inspirational groups and networks that use their strengths and often limited resources to change the status quo. Their actions have raised questioned, stimulated debate, inspired people and form part of a global network that is building just, fairer and more equitable societies.

Kim Bryan and Paul Chatterton are longstanding campaigners involved in direct action around issues such as genetically modified foods, climate change, anti-road and runway building, anti-privatisation struggles and developing anti-capitalist networks. Additional material gratefully received from Dave Morris of Haringey Solidarity Group, Groundwell, Seeds for Change, Turning the Tide, the Ruckus Society, Blockading for Beginners, the Biotic Baking Brigade and Reclaim the Streets.

resources

Books

Abbey, E. (1985). *The Monkey Wrench Gang.* Salt Lake City: Dream Garden Press.

Bari, J. (1994). *Timber Wars.* California: Common Courage Press.

Barlow, Maude (2001). *Blue Gold: The Global Water Crisis and the Commodification of the World's Water Supply.* Rev. edn. Ottawa: Council of Canadians, IFG Committee on the Globalization of Water.

Cockburn, A. and J. St Clair (2000). *Five Days That Shook the World: The Battle for Seattle and Beyond.* London: Verso Books.

Daniels, J. (1989). *Breaking Free. The Adventures of Tin Tin.* London: Attack International.

de Cleyre, V. (undated). *Direct Action,* http://dwardmac.pitzer.edu/Anarchist_Archives/bright/cleyreCW.html

Do or Die (2004). *Voices from the Ecological Resistance.* Vol. 10, Brighton: Do or Die, www.eco-action.org/dod

Evans, Kate (1998). *Copse.* Biddestone, Wiltshire: Orange Dog Productions.

Hindle, Jim (2006). *Nine Miles. Two Years of Anti-roads Protest.* Brighton: Underhill Books.

Katsiaficas, G. and E. Yung (2001). *Battle of Seattle: Debating Corporate Globalization and the WTO.* New York: Soft Skull Press.

Mckay, George (1999). *DIY Culture. Party and Politics in 90s Britain.* London: Verso.

Notes from Nowhere (eds) (2003). *We Are Everywhere: The Irresistible Rise of Global Anti-capitalism.* London: Verso.

Olivera, Oscar (2004). *Cochabamba! Water War in Bolivia.* New York: South End Press.

PGA Women (2001). *Desire for Change: Women on the Frontline of Global Resistance.* London Action Resource Centre, www.nadir.org/nadir/initiativ/agp/gender/desire/desirefor-change.pdf

Solnit, D. (2004). *Globalise Liberation. How to Uproot the System and Build a Better World.* San Francisco: City Lights Books.

Sparrow, R.. (2001). 'Anarchist Politics and Direct Action'. In *Väärin ajateltua: anarckistisia puheenvuoroja herruudettmmasta yhteiskunnasta (Kampus Kustannus, Kopijyvä, Jyväskykä).* Eds T. Ahonen, M. Termonen, T. Tirkkonen, and U. Vehaluoto. 163–75. (in Finnish.)

The Land is Ours (1999). *Beating the Developers. An Activist Guide to the Planning System.* Oxford: The Land is Ours, www.tlio.org.uk/pubs/agp2.html

The Ruckus Society (2003). *Action Planning Training Manual.* Oakland, CA: The Ruckus Society, www.rukus.org

Wall, D. (1999). *Earth First! And the Anti-roads Movement: Radical Environmentalism and Comparative Social Movements.* London: Routledge.

Websites

Campaign resources/networks

Anarchist Black Cross www.abc.org

Anarchists Against the Wall www.squat.net/antiwall

Biotic Baking Brigade www.bioticbakingbrigade.org

Corporate Watch UK www.corporatewatch.org.uk

Corporate Watch USA www.corpwatch.org

Delia Smith's Basic Blockading www.talk.to/delia

Dissent! www.dissent.co.uk

Earth First! UK www.earthfirst.org.uk

Earth First! USA www.earthfirst.org

Genetix Snowball www.fraw.org.uk/gs

Global Calendar of Actions, Protests and Gatherings www.protest.net

International Workers Association www.iwa-ait.org

International Workers of the World www.iww.org

Janitors for Justice www.seiu.org/property/janitors

Legal Defence and Monitoring Group www.ldmg.org.uk

Peoples Global Action www.agp.org

Platform www.platformlondon.org

Rising Tide www.risingtide.co.uk

Road Raging for Beginners www.eco-action.org/rr/

Roadblock www.roadblock.org.uk

School Students Against the War www.ssaw.co.uk

Seeds for Change http://seedsforchange.org.uk/free/resources (Resources for funding,
    planning campaigns, media, legal, training and publicity.)

Shell to Sea http:www.corribsos.com

Smash Edo www.smashedo.org.uk

Subvertising www.adbusters.org

The Land is Ours www.tlio.org.uk

Trident Ploughshares www.tridentploughshares.org

Turning the Tide www.turning-the-tide.org/

Women in Black www.womeninblack.net

conclusion changing our worlds

The Trapese Collective

The contributors to this book have brought alive a world of examples and experiences of how we can do it ourselves, and manage our own lives, collectively without governments, bosses and leaders. These examples may have raised many more questions than they have answered. How can we pull all this together and what does it add up to?

conversations with the future

Imagine each of the chapters in this book are people in your life. They have engaged us in conversations about the future – not a future that we might someday reach, but one they are putting into practice right here in the present. They have presented us with a series of choices about how we can act differently. These acts do require us to take a leap, it's about believing that we have the abilities to change our worlds and being realistic that the world we live in is unsustainable as it is now and neither the market nor governments have the ability or the commitment to tackle the problems we face. This dose of reality may be hard to swallow but there doesn't seem to be an alternative if we want to create just and sustainable lives.

So what points have the chapters made? Talking about sustainability, Andy Goldring commented: 'It doesn't really matter where you start. Follow your curiosity and passion, make it part of your life with practical action and steady learning.' Bryce Gilroy-Scott raised the issue of 'how we will fulfil our basic food, shelter and heat needs in light of the great changes we are facing'. To these ends, he showed us a number of practical ways to make our lives more energy efficient and sustainable (a shower heated by the sun, compost toilet, clean water filtration).

The Seeds for Change Collective asked us to think about how 'we can work with each other rather than seeking to control and command' or be told what to do and suggested that making decisions by consensus can be the bedrock of transforming our world and our relationships with others. Tash Gordon and Becs Griffiths explained how 'we need to change our society in order to improve our health' and how to have more control over our health – through self-help groups and natural remedies. The Trapese Collective discussed 'the need for education where we relearn co-operation and responsibility in order to deal with the crises we face', and presented techniques to empower people and inspire change, including learning our own histories, active learning strategies and action planning. Alice Cutler and Kim Bryan made a call for us to reclaim the land which lies under the concrete. 'We need to redesign our urban and rural space to maximise food production' not only to tackle climate change but also to 'take a step away from the grip of capitalism'. Jennifer Verson suggested that through our actions 'we don't just rehearse the revolution, we practice it everyday', while the Vacuum Cleaner recounted stories of subversive pranksters who pulled amazing stunts on some of the worlds biggest corporations (Starbucks, Nike and Union Carbide.)

Paul Chatterton and Stuart Hodkinson reflected on the 'strong desire and ability people have always shown to determine how and where they live, by collectively managing their own spaces'. A prominent example concerns setting up self-managed social centres and, as Matilda Cavallo discussed, although they involve much work, are essential places to build movements for change. Chekov Feeney explained how people 'don't have to rely on media moguls' as we all have 'the power to describe the world' by building independent media. Mick Fuzz showed us through examples of community newsletters, participatory videos, spoof newspapers and radical film screenings how we can inform and inspire people. Finally, we discussed that 'ordinary people taking direct action, standing up and acting together for their beliefs', has long been the foundation for change and then talked about how to pull a campaign together, whether in your workplace or community or taking on a company, government or policy.

expressing doubts

Although the contributors have made some strong cases, there will inevitably be lingering doubts, criticisms and misunderstandings. It's important to explore some of these here. People often say, *'Things seem OK as we have everything we need for a decent life – a house, car, holidays, clothes and food.'* Demanding radical change and talking about crises may seem distant if we have a comfortable life. But is this sustainable into the future and who else do we affect by leading these lives? We might not often recognise the plight of others who do our work for us around the world – the Bolivian tin miner, the Thai seamstress or the Chinese microchip maker. How do our privileges rest on the suffering of others and what long-term effects do our lifestyle choices have on the planet? We have to face up to the biggest illusion of the twenty-first century that the world has infinite resources and that all of us can lead a high consumption lifestyle. Considering both disturbing climate change and peak oil scenarios which have been presented in this book, all of us need to give up some of our comforts and dependency on consumer lifestyles. Our democracies have become built upon a set of individual rights with the assumption that we are entitled to consume whatever we can afford and without limit. This is especially true in the richer West where we consume a disproportionate amount of global resources, but it is also equally true in rapidly industrialising countries. We are not suggesting a return to the communist era of food rationing or centralised control, where people have to sacrifice everything for the greater good. But these changes need to be made and will be far less difficult if they are undertaken together, and if they allow us to develop more collective notions of rights which take into consideration how our industrial consumer societies impact heavily on others.

A further crucial doubt may concern: '*Could we really run our own communities, make sure we have enough food, purify the water, build our own houses, develop our own media, educate our children?*' It seems difficult, but we do a lot of it already. Many skills already exist amongst us – to teach, look after our health, build and make things, and grow food. The contributors have also shown how we can easily acquire new abilities through skill sharing together. All of us are capable of managing our own lives and becoming experts in what needs to be done. To do this, time and resources need to be freed up by, for example, limiting large salaries at the top, and rethinking work by reducing 9–5 work routines and mindless (call centres), pointless (advertising, excessive accounting and consulting, executives) jobs that dominate our modern economy. We could then focus on the things that really make a difference to people's lives – like more free time for family and friends, good food, and a clean and safe environment.

'*Is this a plan to install a new leadership?*' This book is not a proposal for a new set of leaders to take control, but based on an open-source ethic, a culture that supports and promotes the sharing and collective development of freely available alternatives and solutions to problems. Sharing and collective ownership are direct ways to stop hierarchies and new leaderships or vanguards emerging.

'*So how is this book relevant to my everyday problems like bringing up my kids, work, debt?*' Many may feel defeated and demoralised from the grind of daily life and the messages in this book may seem remote and academic. But it is exactly the opposite. The messages are intended to tackle the root causes of problems associated with work, money and social problems. Strong communities where people communicate openly and support one another can really address some of our everyday issues. Parenting clubs, gardening classes, sharing our costs and resources, supporting those who are ill, building community media and spaces – these are the things that can respond to some of the daily problems we face. And as the chapters on health stressed, the things that most affect our well-being are the social conditions we experience and the type of society we live in. For people to actually do it themselves there also has to be enough time. In a world where most people are working flat out to keep on top of debts this is a major issue. One thing we will have to learn is how to resist the multi-billion pound advertising onslaught and free ourselves from the idea that having the latest fad and fashion will make us happy. We will slowly have to redefine leisure and work, and look for ways to redirect resources, finding ways to live cheaply, cutting out the middlemen making a quick profit, and being resourceful and bold about our lifestyle decisions. Do we really need to spend what little spare money and time we have shopping for more clothes and more things?

'We need some people in control, otherwise society will break down.' Some may feel that society works very well and that we need police and governments, especially in a world that is increasingly global and complex. There are legitimate fears that if we got rid of governments, society would be taken over by terrorists and descend into violent rival gangs competing for resources. Every society or community needs mechanisms for organising and reproducing itself. But these can be based on voluntary associations, co-operation and decentralised, consensus decision making. Resource competition would also be reduced if they were allocated more equally. There will always be people who dominate, but the majority of us co-operate with each other peacefully. In Chapter 1, Starhawk gave the example of relief work after Hurricane Katrina in New Orleans where activists in the local community instinctively co-operated and helped each other. The group Common Ground was set up which undertook neighbourhood protection garbage pickup, distribution of relief supplies and a free medical clinic. The Common Ground collective continues to supply much needed help and support in rebuilding sustainable communities in the region.

'Will groups that are intensely divided just simply start working together?' This book is not naïvely proposing that there are not people who operate from a basis of selfishness, racism and hostility. These problems create a culture of fear (like not trusting your neighbour or anyone that is different) and this is only exacerbated by our individual-istic societies. There will always be differences between people and groups, and debate and discussion are important starting points to explore, understand and resolve these differences. There are many ways to build community and challenge these notions of prejudice. Popular education can be one way to bring people together on an equal basis and to learn co-operation. As the Seeds for Change Collective showed in Chapters 3 and 4, the key is relearning how to make decisions together, without relying on experts or leaders, looking for consensus rather than imposing our views on others. Chapter 13 reflected on the importance of creating social spaces that are open to everyone for people to meet, work, organise and have fun with each other. Planting community gardens was suggested in Chapter 10 as another potential catalyst for community development. While there is currently only a smattering of such projects, they have great potential if they spread to every street and village. As one of Aesop's fables said, 'united we stand, divided we fall'.

'*Won't self-management mean a massive amount of work that will take a long time to realise?*' Self-managed, horizontal politics does mean extra meetings and time spent finding agreements between different viewpoints. Keeping track of what is decided and resolving conflicts requires commitment, but as understanding and trust is developed the time spent in meetings begins to pay off. The extra time this takes is nothing compared to the amount of time saved from, for example, reduced commuting, shopping or some jobs. Change will be slow and necessarily so in order that everyone has time to participate. Realistically, larger-scale transformation is still a long way off. But this book is part of preparing for that possibility, laying the ground, learning together, building solidarity and developing our own solutions. Once we are alive to the problems we face and aware of the possibilities to tackle them, then there's no going back. It is about learning the nature of the system we are challenging, how it works and what tactics we need to build alternatives which are parallel to it.

making the leap

There are some very alarming possibilities and even probabilities as to what is unfolding. As we write this book the conflict and war in the Middle East is escalating daily; we are feeling the effects of global climate change as anticyclones and higher temperatures become more frequent; and the democratic deficit and open hypocrisy of Western government is exemplified by falling numbers of voters and leaders who ignore our demands for peace, change and public services. Our own creativity and confidence becomes further squashed and undervalued as a result, as is our ability and desire to create alternatives. But options and solutions are spreading around the world and, despite these worst case scenarios, there are many movements and people that highlight and represent the hope and optimism of these global networks.

Our strategy should be not only to confront empire, but to lay siege to it. To deprive it of oxygen. To shame it. To mock it. With our art, our music, our literature, our stubbornness, our joy, our brilliance, our sheer relentlessness – and our ability to tell our own stories. Stories that are different from the ones we're being brainwashed to believe ... another world is possible, she is on her way and on a quiet day, I can hear her breathing. (Arundhati Roy at the World Social Forum, Porto Alegre 2003)

Whilst the problems of creating radical and meaningful social change through empowering communities to do it themselves may seem initially overwhelming, there is much to learn and share from the successes and failures of popular movements. Struggle and resistance have been an ongoing part of our collective human histories – be it against colonialism, slavery, feudalism, patriarchy or, more recently, global institutions and their elites. These struggles do not occur in isolation but as part of a continuum through time and show that collective resistance can succeed in challenging oppression.

Taken on their own, the ideas and examples in this book might seem a small contribution to a huge problem. But, in fact by talking and acting on these we are involved in the debate about the future, how the twenty-first century will unfold. These ideas are recipes to experiment with on your own terms, in your own communities – starting points not blueprints. The answers will really only come out through living it and doing it – and making sure we keep talking and figuring it out as we go along. It's not anyone's job to convince people of the worth of particular ways of organising life. The message of this book is that everyone has the answers for themselves, given the time, patience and skills to find them. People will be pulled towards what we are saying if they see workable alternatives that actually improve their lives, and pushed towards them as the problems associated with the current way we organise our societies mount. Getting involved in managing our own lives and communities should not be a burden. Yes there's going to be responsibility, but there's also going to be creativity, freedom and fun.

Considering the uncertainty, it is normal to be nervous or unclear. There are only a few workable examples to compare and believing in their potential involves a leap of faith. But perpetuating the present, with all its drudgery, unfulfilled potential, pointless jobs, massive and unsustainable movement of goods, impoverished communities, poor housing, ecological crisis points, human rights abuses, and lack of meaningful democracy is no alternative, except for a small elite who designed and benefit from this system. There is no side stepping the crucial issues our contributors have raised. It is up to us to urgently respond together – elected leaders, corporations and the mechanisms of the market are, by design, unable to do so.

These conversations have no end – the answers come through talking, doing and experimenting together. And, in the same way, this book has no fixed conclusions.

We invite you to take these chapters and share, improve, diffuse, plunder, rewrite and abuse the ideas you have found in these pages; they are yours to use.

As a friend of ours, Claire Fauset, says:

> We can't tell you anything new.
> In all these millennia of human existence
> there can only be a few new ideas to be thought through.
> So do we treat them like rare commodities?
> Or do we re-use and recycle them?
> Pile our public spaces high with shared ideas beyond anyone's imagining.
> The revolution will be plagiarised!
> The revolution will not happen if ideas are corporatised.
> So read this book and use it
> with your own ending and for your own ends ...

index

Compiled by Sue Carlton